How to Generate a Genuine Love of Mathematics

Problem Solving in Mathematics and Beyond

Print ISSN: 2591-7234
Online ISSN: 2591-7242

There are countless applications that would be considered problem solving in mathematics and beyond. One could even argue that most of mathematics in one way or another involves solving problems. However, this series is intended to be of interest to the general audience with the sole purpose of demonstrating the power and beauty of mathematics through clever problem-solving experiences.

Each of the books will be aimed at the general audience, which implies that the writing level will be such that it will not be engulfed in technical language — rather the language will be simple everyday language so that the focus can remain on the content and not be distracted by unnecessarily sophiscated language. Again, the primary purpose of this series is to approach the topic of mathematics problem-solving in a most appealing and attractive way in order to win more of the general public to appreciate this most important subject rather than to fear it. At the same time we expect that professionals in the scientific community will also find these books attractive, as they will provide many entertaining surprises for the unsuspecting reader.

Published

Vol. 45 *How to Generate a Genuine Love of Mathematics*
by Alfred S Posamentier

Vol. 44 *Revolutions in Mathematics*
by Arthur D Kramer and Alfred S Posamentier

Vol. 43 *Mathematical Puzzles and Curiosities*
by Ivo Fagundes David de Oliveira, Tanya Khovanova and Yogev Shpilman

Vol. 42 *Learning Geometry Through Problem Solving*
by Boris Pritsker

For the complete list of volumes in this series, please visit www.worldscientific.com/series/psmb

Problem Solving in Mathematics and Beyond | Volume 45

How to Generate a Genuine Love of Mathematics

Alfred S. Posamentier
The City University of New York, USA

World Scientific

NEW JERSEY • LONDON • SINGAPORE • BEIJING • SHANGHAI • HONG KONG • TAIPEI • CHENNAI • TOKYO

Published by

World Scientific Publishing Co. Pte. Ltd.
5 Toh Tuck Link, Singapore 596224
USA office: 27 Warren Street, Suite 401-402, Hackensack, NJ 07601
UK office: 57 Shelton Street, Covent Garden, London WC2H 9HE

Library of Congress Cataloging-in-Publication Data
Names: Posamentier, Alfred S. author
Title: How to generate a genuine love of mathematics / author,
Alfred S Posamentier (The City University of New York, USA).
Description: Hackensack, NJ ; Covent Garden, London : World Scientific
Publishing Co. Pte. Ltd., [2026] | Series: Problem solving in
mathematics and beyond, 2591-7234 ; volume 45 | Includes index.
Identifiers: LCCN 2026012577 | ISBN 9789819830411 hardcover |
ISBN 9789819830893 paperback | ISBN 9789819830428 ebook |
ISBN 9789819830435 ebook other
Subjects: LCSH: Mathematics--Study and teaching
Classification: LCC QA11.2 .P6388 2026
LC record available at https://lccn.loc.gov/2026012577

British Library Cataloguing-in-Publication Data
A catalogue record for this book is available from the British Library.

For any available supplementary material, please visit
https://www.worldscientific.com/worldscibooks/10.1142/14798#t=suppl

Desk Editors: Kannan Krishnan/Gabriel Rawlinson

Typeset by Stallion Press
Email: enquiries@stallionpress.com

About the Author

Alfred S. Posamentier was most recently a Distinguished Lecturer at New York City College of Technology of the City University of New York. Prior to that he was Executive Director for Internationalization and Funded Programs at Long Island University, New York. This was preceded by 5 years as Dean of the School of Education and Professor of Mathematics Education at Mercy University, New York. Before that he was for 40 years at the City College of the City University of New York, at which he is now Professor Emeritus of Mathematics Education and Dean Emeritus of the School of Education. He is the author and co-author of more than 90 mathematics books for teachers, secondary and elementary school students, as well as the general readership. Dr. Posamentier is also a frequent commentator in newspapers and journals on topics related to education and mathematics.

After completing his B.A. degree in mathematics at Hunter College of the City University of New York, he took a position as a teacher of mathematics at Theodore Roosevelt High School (Bronx, New York), where he focused his attention on improving the students' problem-solving skills and at the same time enriching their instruction far beyond what the traditional textbooks offered. During his six-year tenure there, he also developed the school's first mathematics teams (both at the junior and senior level). He is still involved in working with mathematics teachers and supervisors, nationally and internationally, to help them maximize their effectiveness.

Immediately upon joining the faculty of the City College of New York in 1970 (after having received his master's degree there in 1966), he began to develop in-service courses for secondary school mathematics teachers, including such special areas as recreational mathematics and problem solving in mathematics. As Dean of the City College School of Education for 10 years, his scope of interest covered the full gamut of educational issues. During his tenure as Dean, he took the School of Education from the bottom of the New York State rankings to the top with a perfect NCATE accreditation assessment in 2009. He also raised more than $12,000,000 from the private sector for educational innovative programs. Posamentier repeated this successful transition at Mercy University, where he enabled it to become the only college to have received both NCATE and TEAC accreditation simultaneously.

In 1973, Dr. Posamentier received his Ph.D. from Fordham University (New York) in mathematics education, and he has since extended his reputation in mathematics education to Europe. He has been visiting professor at several European universities in Austria, England, Germany, the Czech Republic, Turkey and Poland. In 1990, he was Fulbright Professor at the University of Vienna.

In 1989 he was awarded an Honorary Fellow position at the South Bank University (London, England). In recognition of his outstanding teaching, the City College Alumni Association named him Educator of the Year in 1994, and in 2009. New York City had the day, May 1, 1994, named in his honor by the President of the New York City Council. In 1994, he was also awarded the *Grosse Ehrenzeichen für Verdienste um die Republik Österreich* (Grand Medal of Honor from the Republic of Austria), and in 1999, upon approval of Parliament, the President of the Republic of Austria awarded him the title of University Professor of Austria. In 2003, he was awarded the title of *Ehrenbürgerschaft* (Honorary Fellow) of the Vienna University of Technology, and in 2004 he was awarded the *Österreichisches Ehrenkreuz für*

Wissenschaft & Kunst 1. Klasse (Austrian Cross of Honor for Arts and Science, First Class) from the President of the Republic of Austria. In 2005, he was inducted into the Hunter College Alumni Hall of Fame, and in 2006 he was awarded the prestigious Townsend Harris Medal by the City College Alumni Association. He was inducted into the New York State Mathematics Educator's Hall of Fame in 2009, and in 2010 he was awarded the coveted Christian-Peter-Beuth Prize from the Technische Fachhochschule—Berlin. In 2017, Posamentier was awarded *summa cum laude nemine discrepante* by the Fundacion Sebastian, A.C., Mexico City, Mexico.

He has taken on numerous important leadership positions in mathematics education locally. He was a member of the New York State Education Commissioner's Blue Ribbon Panel on the Math-A Regents Exams, and the Commissioner's Mathematics Standards Committee, which redefined the Mathematics Standards for New York State, and he also served on the New York City Schools' Chancellor's Math Advisory Panel.

Dr. Posamentier is still a leading commentator on educational issues and continues his long-time passion of seeking ways to make mathematics interesting to teachers, students and the general public—as can be seen from some of his more recent books.

For more information and a list of his publications see: https://en.wikipedia.org/wiki/Alfred_S._Posamentier.

Contents

Introduction

Most people are proud to admit they did not do well in mathematics during their school days. No other subject in the school curriculum is more hated by the general population than math. Frankly, it's upsetting to hear that one of the most important and fundamental fields of study, taught from preschool and kindergarten through high school, receives such a negative assessment from so many. Typically, students enjoy subjects that are taught well. Unfortunately, due to the stresses of grades and standardized tests, mathematics teachers have prioritized preparing students to do well on mathematics exams. This has frequently led to "teaching to the test," which leaves little time for teachers to digress from the standard curriculum and introduce topics and ideas that can make mathematics far more interesting. Investing time to briefly deviate from the curriculum can spark a student's genuine interest in math, thereby motivating the student to pursue an understanding of the subject that extends beyond just passing the test.

Sadly, many adults didn't have pleasant experiences in mathematics courses throughout school. This book presents basic mathematical topics that should inspire awe of the subject and thereby, positively change the reader's perception of mathematics. When adults rearing children begin to have a more favorable view of mathematics—one that escaped them during their own education—the next generation could benefit from at-home encouragement. This will be just the first step to improving understanding of and performance in mathematics.

Very little background mathematics is necessary for the reader of this book to fully comprehend and, more importantly, appreciate the topics covered. Arithmetic, which today is typically performed on a computer, can sometimes be performed more quickly mentally, provided one knows some uncommon methods. Furthermore, we live in a geometric world, where there are untold aspects of geometry that are not only generally interesting but also helpful in navigating today's thinking. However, the first thing we learn about in mathematics is numbers. There is so much beauty and amazement in our number system that is unfortunately absent from the elementary school curriculum. We also offer a chapter that is intended to sharpen logical thinking, which is not only useful in mathematics but also in everyday life. Join us now as we appreciate the amazing relationships between numbers.

Chapter 1

Entertainment with Number Peculiarities

To introduce the uninitiated to the beauty of mathematics it may be best to begin from an elementary standpoint, where we can unveil the hidden wonders in mathematics. Our journey through the amazing aspects of mathematics will start with the curiosities, coincidences, and overall marvelous relationships involving only number properties that are generally not known by most people. As we all have a certain facility with numbers, we will be impressed with the miraculous secrets that numbers keep from the general public. Once one becomes familiar with these aspects of simple number relationships, one will want to pursue other mysteries. That should serve as a good foundation for a genuine love for mathematics that seems to have bypassed many folks throughout their schooling.

This chapter will investigate the properties, characteristics, and relationships between and among numbers. We start with the number 1 and selectively increase numbers until we get to huge, potentially intimidating numbers that also carry unexpected properties and relationships. The numbers are presented in a way that one can appreciate the wonders and beauty that they represent. No mathematics beyond arithmetic is needed to understand these curiosities. Of course, one could use a calculator to verify some of the exposed relationships. As this book was written in the year 2025, you might look forward to the amazing characteristics of this year's number (2025) when you reach it numerically. Not only is it a perfect square ($2025 = 45^2$), but only a reader over 90 years old will have experienced 1936 (which is 44^2), the last year prior to 2025 that is a perfect square. Most of us will not see the next perfect square year: 2116 (which is 46^2). Join us now as we begin with the smallest natural number, 1, and progress through other remarkable numbers sequentially.

Irrepressible Number 1

Naturally, the number 1 is basic in any number system and we hardly think much about it. However, the number 1 appears in many surprising circumstances. We present one way to consider a "magical" property of the number 1 that has baffled mathematicians for many years. To this day, no one knows why 1 posseses this property. It is known as the *Collatz conjecture*, and was named after the German mathematician Lothar Collatz (1910–1990). The process we are about to engage in begins with selecting any number and always ends with the number 1. We begin with the following two rules as you work with any *arbitrarily* selected number.

- *If a given number is odd, then multiply it by 3 and add 1.*
- *If the number is even, then simply divide the number by 2.*
- *Continue this process until you reach the number 1.*
- *Regardless of the number originally selected, the result will always be 1.*

Let's try it for the *arbitrarily selected* number **12.**
Since 12 is even, therefore, divide by 2 to get 6.
Since 6 is also even so we divide by 2 to get 3.
Since 3 is odd, therefore, multiply by 3 and add 1 to get 10.
The number 10 is even, so we simply divide by 2 to get 5.
Since 5 is odd, we multiply by 3 and add 1 to get 16.
Since 16 is even so we divide by 2 to get 8.
Since 8 is even so we divide by 2 to get 4.

Since 4 is even so we divide by 2 to get 2.

Since 2 is even so we divide by 2 to get 1, which the Collatz conjecture predicted.

Regardless of which number we begin with (in the above example we started with 12) we will eventually get to 1. This is truly remarkable! Had we started with 17 as our arbitrarily-selected starting number, we would have required 13 steps to reach the number 1. Starting with the number 43 will require 27 steps to reach the number 1.

Does this really work for all numbers? This is a question that has concerned mathematicians since the 1930s, and to date no answer has been found, despite monetary rewards having been offered for a proof of this conjecture. Most recently (using computers) this problem, known in the literature as the "$3n + 1$ Problem," has been shown to be true for the numbers up to $10^{18} - 1$. The longest progression of steps, so far established, is for the number 931,386,509,544,713,451, which requires 2,283 steps to reach the number 1. So, you can see how the harmless number 1 is omnipresent.

There are, however, "tricks" to generate the number 1 that we can justify using algebra. For example, select a random number and then double it and add 2 to the result and then take half and subtract the number you began with. The result will always be 1. Let's try this with the number 18 as our beginning number. We double 18 to get 36, whereupon we add 2 to get 38, then take half, which is 19, and then subtract the original number, 18, to get $19 - 18 = 1$. As opposed to the Collatz conjecture, this procedure can be easily justified algebraically as follows:

$$(2x+2)\cdot\frac{1}{2}-x=x+1-x=1.$$

The Ubiquitous Number 2

The number 2 seems to be omnipresent as a member of the number system. Here are some unusual powers and properties exhibited by the number 2.

- The number 2 is the only number whose self-sum and self-product is the same, that is, $2 + 2 = 4$, and $2 \times 2 = 4$.

- The number 2 is the first prime number, since it has exactly 2 divisors: 1 and 2.
- The number 2 is the only even prime number.
- The number 2 is the only prime followed consecutively by another prime number, namely, 3.
- Powers of 2 appear more frequently in mathematics than powers of any other numbers.
- Positive integers can be expressed as the sum of 2 or more consecutive integers, but only if they are not a power of 2.
- The number 2 is the only prime number whose cube is a sum of 2 consecutive prime numbers, that is, $2^3 = 8 = 3 + 5$.
- The number 2 is the only prime number that is not the difference of 2 square numbers.
- The number 2 is the only prime number that can be represented in the form $n^n + n$, which is true when $n = 1$.
- Every number can be expressed in a unique way as the sum of powers of 2. For example: $59 = 32 + 16 + 8 + 2 + 1$.
- The famous French mathematician Pierre de Fermat (1607–1665) discovered that if n is a prime number, then n will always be a factor of $2^n - 2$. For example, when $n = 5$ then $2^5 - 2 = 30$, of which 5 is a factor. Also, when n is a prime number, then $2^n - 1$ will always be a prime number as well. For example, when $n = 7$, we have $2^7 - 1 = 127$, which is a prime number.

Here is a rather strange curiosity: 2^{168} is the largest power of 2 whose decimal expansion does not contain the digit 2. That is, 2^{168} = 374144419156711147060143317175368453031918731001856.

Thus, we can appreciate the special role that the number 2 has in both mathematics and beyond.

Surprising Characteristics of the Number 3

As we look at the number 3, we will search for unusual characteristics and relationships. The number 3 is the only number equal to the sum of all the preceding positive integers: $1 + 2 = 3$. Furthermore, a unique circle is determined by 3 non-collinear points. The number 3 is the only prime number that yields a prime number when 2 is added to its square, which is $3^2+2=11$.

The number 3 also has a strange property that is difficult to explain but attractive to experience. When you take 3 to any power the result is a number where the sum of the digits is 9 or a multiple of 9. To convince you that this is correct we provide a few examples:

$$3^2 = 9,\ 3^3 = 27,\ 3^4 = 81, \text{ and } 3^5 = 243.$$

Notice that for each power the sums of digits are equal to 9.

The next powers of 3 have a digit-sums which are multiples of 9:

$$3^6 = 729,\ 3^7 = 2187,\ 3^8 = 6561,\ 3^{17} = 129{,}140{,}163.$$

The famous German mathematician Carl Friedrich Gauss (1777–1855) showed that every integer is the sum of at most 3 triangular numbers, which are numbers that resemble points that can be arranged in triangular form, such as the numbers 1, 3, 6, 10, 15, 21, 28, etc. For example, the number 39 can be expressed as the sum of three triangular numbers as $39 = 28+10+1$. (See page 94.)

A Curious Characteristic of the Number 4

The peculiarity of numbers is always an attraction to mathematics. A noteworthy feature of the number 4 occurs when we take any number, take the sum of the squares of its digits, and continue this process with each resulting number. We will either end up with the number 4 or occasionally, the number 1. For example, let's begin with the number 5 and continually take the sum of the squares of its digits until we notice a pattern: $5 \to 25 \to 29 \to 85 \to 89 \to 145 \to 42 \to 20 \to 4$. When we continue this with the number 4, notice what results: $4 \to 16 \to 37 \to 58 \to 89 \to 145 \to 42 \to 20 \to 4$. Thus, we see that the cycle continues whenever it reaches the number 4. When the number reaches the number 1 the cycle ends, whereas when it reaches the number 4 it continues on indefinitely. You will see later as we progress through our numerical journey that when an initially selected number processed in this way ends in a 1, we call that initial number a *happy number* (see page 100). When it ends in a 4, we refer to this initial number as an *unhappy number*, perhaps since it goes on in the never-ending cycle.

In number theory, we have the four-square theorem, which was proved in 1770 by the Italian mathematician Joseph Louis Lagrange (1736–1813) and states that every natural number can be expressed as the sum of four integer squares. Here are a few examples to illustrate the theorem:

$$2 = 1^2 + 1^2 + 0^2 + 0^2$$

$$31 = 1^2 + 1^2 + 2^2 + 5^2$$

$$53 = 6^2 + 4^2 + 1^2 + 0^2$$

$$310 = 1^2 + 2^2 + 4^2 + 17^2$$

Pierre de Fermat discovered a unique feature of the number 4, which is that every prime number one greater than a multiple of 4 can be expressed uniquely as a sum of two squares. For example, the number 17 is one greater than 16, a multiple of 4, and it can be expressed as $1^2 + 4^2$. As another example, the number 45 (44 + 1) can be expressed as $3^2 + 6^2 = 45$. This principle can be extended to read that *no* number *one less than* a multiple of 4 can be expressed as a sum of two squares.

Just for entertainment and a challenge, we can generate the numbers 1 through 100 using only four 4s each time.

$$0 = 44 - 44 = \frac{4}{4} - \frac{4}{4}$$

$$1 = \frac{4+4}{4+4} = \frac{\sqrt{44}}{\sqrt{44}} = \frac{4+4-4}{4}$$

$$2 = \frac{4 \cdot 4}{4+4} = \frac{4-4}{4} + \sqrt{4} = \frac{4}{4} + \frac{4}{4}$$

$$3 = \frac{4+4+4}{4} = \sqrt{4} + \sqrt{4} - \frac{4}{4} = \frac{4 \cdot 4 - 4}{4} = 4 - 4^{4-4}$$

$$4=\frac{4-4}{4}+4=\frac{\sqrt{4\cdot 4}\cdot 4}{4}=(4-4)\cdot 4+4$$

$$5=\frac{4\cdot 4+4}{4}$$

$$6=\frac{4+4}{4}+4=\frac{4\sqrt{4}}{4}+4$$

$$7=\frac{44}{4}-4=\sqrt{4}+4+\frac{4}{4}=(4+4)-\frac{4}{4}$$

$$8=4\cdot 4-4-4=\frac{4(4+4)}{4}=4+4+4-4$$

$$9=\frac{44}{4}-\sqrt{4}=4\sqrt{4}+\frac{4}{4}=\frac{4}{4}+4+4$$

$$10=4+4+4-\sqrt{4}=\frac{44-4}{4}$$

$$11=\frac{4}{4}+\frac{4}{.4}=\frac{44}{\sqrt{4}+\sqrt{4}}$$

$$12=\frac{4\cdot 4}{\sqrt{4}}+4=4\cdot 4-\sqrt{4}-\sqrt{4}=\frac{44+4}{4}$$

$$13=\frac{44}{4}+\sqrt{4}$$

$$14=4\cdot 4-4+\sqrt{4}=4+4+4+\sqrt{4}=\frac{4!}{4+4+4}=4!-\left(4+4+\sqrt{4}\right)$$

$$15=\frac{44}{4}+4=\frac{\sqrt{4}+\sqrt{4}+\sqrt{4}}{.4}$$

$$16=4\cdot 4-4+4=\frac{4\cdot 4\cdot 4}{4}=4+4+4+4$$

$$17=4\cdot 4+\frac{4}{4}$$

$$18=\frac{44}{\sqrt{4}}-4=4\cdot 4+4-\sqrt{4}=4\cdot 4+\frac{4}{\sqrt{4}}=\frac{4!+4!+4!}{4}$$

$$19=\frac{4+\sqrt{4}}{.4}+4=4!-4-\frac{4}{4}$$

$$20=4\cdot 4+\sqrt{4}+\sqrt{4}=\left(4+\frac{4}{4}\right)\cdot 4$$

$$21=4!-4+\frac{4}{4}=4!4\cdot 4+4+\sqrt{4}$$

$$22=\frac{4}{4}(4!)-\sqrt{4}=4!-\frac{((4+4)/4)}{4}=\frac{44}{4}\cdot\sqrt{4}=-4+\frac{4}{4}$$

$$23=4!-\sqrt{4}+\frac{4}{4}=4!-4^{4-4}$$

$$24=4\cdot 4+4+4$$

$$25=4!-\sqrt{4}+\frac{4}{4}=4!+\sqrt{(4+4-4)}=\left(4+\frac{4}{4}\right)^{\sqrt{4}}$$

$$26=\frac{4}{4}(4!)+\sqrt{4}=4!+\sqrt{4+4-4}=4+\frac{44}{\sqrt{4}}$$

$$27=4!+4-\frac{4}{4}$$

$$28=(4+4)\cdot 4-4=44-4\cdot 4$$

$$29=4!+4+\frac{4}{4}$$

$$30=4!+4+4-\sqrt{4}$$

$$31=\frac{((4+\sqrt{4})!+4!)}{4!}$$

$$32=(4\cdot 4)+(4\cdot 4)$$

$$33=4!+4+\frac{\sqrt{4}}{.4}$$

$$34=\left(4\cdot 4\cdot\sqrt{4}\right)+\sqrt{4}=4!+\left(\frac{4!}{4}\right)+4=\sqrt{(4^4)}\cdot\sqrt{4}+\sqrt{4}$$

$$35=4!+\frac{44}{4}$$

$$36=(4+4)\cdot 4+4=44-4-4$$

$$37=4!+\frac{\left(4!+\sqrt{4}\right)}{\sqrt{4}}$$

$$38=44-\frac{4!}{4}$$

$$39=4!+\frac{4!}{4\cdot .4}$$

$$40=(4!-4)+(4!-4)=4\cdot\left(4+4+\sqrt{4}\right)$$

$$41=\frac{4!+\sqrt{4}}{.4}-4!$$

$$42=44-4+\sqrt{4}=(4!+4!)-\frac{4!}{4}$$

$$43=44-\left(\frac{4}{4}\right)$$

$$44=44+4-4$$

$$45=44+\frac{4}{4}$$

$$46=44+4-\sqrt{4}=(4!+4!)-\left(\frac{4}{\sqrt{4}}\right)$$

$$47-(4!+4!)\quad\begin{matrix}4\\4\end{matrix}$$

$$48=(4\cdot 4-4)\cdot 4=4\cdot(4+4+4)$$

$$49=(4!+4!)+\frac{4}{4}$$

$$50=44+\left(\frac{4!}{4}\right)=44+4+\sqrt{4}$$

$$51=\frac{(4!-4+.4)}{.4}$$

$$52=44+4+4$$

$$53=4!+4!+\frac{\sqrt{4}}{.4}$$

$$54 = 4! + 4! + 4 + \sqrt{4}$$

$$55 = \frac{(4! - 4 + \sqrt{4})}{.4}$$

$$56 = 4! + 4! + 4 + 4 = 4 \cdot (4 \cdot 4 - \sqrt{4})$$

$$57 = \left(\frac{4! - \sqrt{4}}{.4}\right) + \sqrt{4}$$

$$58 = ((4! + 4) \cdot \sqrt{4}) + \sqrt{4} = 4! + 4! + \frac{4}{.4}$$

$$59 = \frac{(4! - \sqrt{4})}{.4} + 4 = \frac{4!}{.4} - \frac{4}{4}$$

$$60 = 4 \cdot 4 \cdot 4 - 4 = \frac{4^4}{4} - 4 = 44 + 4 \cdot 4$$

$$61 = \frac{(4! + \sqrt{4})}{.4} - 4 = \frac{4!}{.4} + \frac{4}{4}$$

$$62 = 4 \cdot 4 \cdot 4 - \sqrt{4}$$

$$63 = \frac{(4^4 - 4)}{4}$$

$$64 = 4\sqrt{4} \cdot 4\sqrt{4} = 4 \cdot (4! - 4 - 4) = (4 + 4) \cdot (4 + 4)$$

$$65 = \frac{(4^4 + 4)}{4}$$

$$66 = 4 \cdot 4 \cdot 4 + \sqrt{4}$$

$$67 = \frac{4! + \sqrt{4}}{.4} + \sqrt{4}$$

$$68 = 4 \cdot 4 \cdot 4 + 4 = \frac{4^4}{4} + 4$$

$$69 = \left(\frac{4! + \sqrt{4}}{.4}\right) + 4$$

$$70 = \frac{(4 + 4)!}{4! \cdot 4!} = 44 + 4! + \sqrt{4}$$

$$71=\frac{4!+4.4}{.4}$$

$$72=44+4!+4=4\cdot\left(4\cdot 4+\sqrt{4}\right)$$

$$73=\frac{4!\cdot\sqrt{4}+\sqrt{.4}}{\sqrt{.4}}$$

$$74=4!+4!+4!+\sqrt{4}$$

$$75=\frac{4!+4+\sqrt{4}}{.4}$$

$$76=(4!+4!+4!)+4$$

$$77=\left(\frac{4}{.4}\right)^{\sqrt{4}}-4$$

$$78=4\cdot(4!-4)-\sqrt{4}$$

$$79=4!+\frac{4!-\sqrt{4}}{.4}$$

$$80=(4\cdot 4+4)\cdot 4$$

$$81=\left(4-\left(\frac{4}{4}\right)\right)^{4}=\left(\frac{4!}{4\sqrt{4}}\right)^{4}$$

$$82=4\,(4!-4)+\sqrt{4}$$

$$83=\frac{4!-.4}{.4}+4!$$

$$84=44\cdot\sqrt{4}-4$$

$$85=\frac{\frac{4!+4}{.4}}{.4}$$

$$86=44\cdot\sqrt{4}-\sqrt{4}$$

$$87=4\cdot 4!-\frac{4}{.4}=44\sqrt{4}-i^{4}$$

$$88=4\cdot 4!-4-4=44+44$$

$$89=4!+\frac{4!+\sqrt{4}}{.4}$$

$$90 = 4 \cdot 4! - 4 - \sqrt{4} = 44 \cdot \sqrt{4} + \sqrt{4}$$

$$91 = 4 \cdot 4! - \frac{-\sqrt{4}}{.4}$$

$$92 = 4 \cdot 4! - \sqrt{4} - \sqrt{4} = 44 \cdot \sqrt{4} + 4$$

$$93 = 4 \cdot 4! - \frac{4}{.4}$$

$$94 = 4 \cdot 4! + \sqrt{4} - 4$$

$$95 = 4 \cdot 4! - \frac{4}{4}$$

$$96 = 4 \cdot 4! + 4 - 4 = 4! + 4! + 4! + 4!$$

$$97 = 4 \cdot 4! + \frac{4}{4}$$

$$98 = 4 \cdot 4! + 4 - \sqrt{4}$$

$$99 = \frac{44}{.44} = 4 \cdot 4! + \frac{\sqrt{4}}{\sqrt{.04}} = \frac{4}{4\%} - \frac{4}{4}$$

$$100 = 4 \cdot 4! + \sqrt{4} + \sqrt{4} = \left(\frac{4}{.4}\right) \cdot \left(\frac{4}{.4}\right) = \frac{44}{.44}$$

Consider the First Perfect Number 6

With the number 6, we encounter the first of a set of numbers known as "*perfect numbers*," which are numbers that are equal to the sum of their proper divisors: 6 = 1 + 2 + 3. The first few perfect numbers are 6, 28, 496, 8128, 33550336, 8589869056, 137438691328, and 2305843008139952128. (See page 87.)

The number 6 is unique in that it's the only number that can be expressed as the sum and product of the same three numbers, namely, $6 = 1 \times 2 \times 3 = 1 + 2 + 3$.

Throughout this book, we will frequently refer to "*prime numbers*," which are numbers that have exactly two factors: 1 and the number itself. (See page 87.) For example, 5 is a prime number, since it has only two factors, namely, 1 and 5, whereas the number 6 is not a prime number since it has more than two factors, namely, 1, 2, 3 and 6. A number which is not a prime number is called a *composite number*. The first few prime numbers are 2, 3, 5, 7, 11, 13, 17, 19, and 23. Yet, in the realm of prime numbers, the number 6 also has a unique characteristic. Every prime number greater than 3 can be expressed in the form of $6n \pm 1$. For example, here are a few prime numbers expressed in this fashion: $37 = (6 \times 6) - 1$, $61 = (6 \times 10) + 1$, and $71 = (6 \times 12) - 1$.

We will end our celebration of the number 6 with a cute little number trick that will always result in the number 6. We begin by taking any three consecutive numbers, where the largest is divisible by 3. We add the numbers, then add the digits of that sum, and continue this process by taking the sum of the digits of the previous result. When a single digit is reached, you will then find that the result will always be 6. For example, to dramatize our technique, let's take three consecutive extremely large numbers and add them:

$$356{,}842{,}780 + 356{,}842{,}781 + 356{,}842{,}782 = 1{,}070{,}528{,}343.$$

Now we find the sum of those digits:

$$1 + 0 + 7 + 0 + 5 + 2 + 8 + 3 + 4 + 3 = 33.$$

We take the sum of the digits again and find $3 + 3 = 6$, which was our desired result.

The Number 7 Can Generate Some Beauty

Having just been introduced to prime numbers, we notice that the number 7 is a prime number, since its only prime factors are 1 and 7. Furthermore, the number 7 also provides us with some arithmetic curiosities, such as when we divide 999,999 by 7, we get 142,857. At first glance, this does not appear to be anything special.

However, if you consider the repeating decimal value of $\frac{1}{7} = 0.142857\mathbf{142857}142857\mathbf{142857}\ldots$,= which is quite spectacular. This can be taken a step further when we consider the following:

$$\frac{2}{7} = 0.285714\underline{\mathbf{285714}}285714\underline{\mathbf{285714}}285714$$

$$\frac{3}{7} = 0.428571\underline{\mathbf{428571}}428571\underline{\mathbf{428571}}428571$$

$$\frac{4}{7} = 0.5714285\underline{714285}714285\underline{714285}71428$$

$$\frac{5}{7} = 0.7142857\underline{142857}142857\underline{142857}14285$$

$$\frac{6}{7} = 0.8571428\underline{571428}571428\underline{571428}57142$$

Notice how the digits repeat themselves in order, each time beginning with a different initial digit.

Another unique property of the number 7 is that the sum of seven consecutive prime numbers beginning with the number 7 is equal to 7 times the 7th prime number, which is 17. That is,

$$7 + 11 + 13 + 17 + 19 + 23 + 29 = 119 = 7 \times 17.$$

We can also create very unusual patterns from powers of 7. Consider the number 14,459,929 and find the sum of each of the digits taken to the seventh power. Amazingly, we arrive at the same number: $1^7 + 4^7 + 4^7 + 5^7 + 9^7 + 9^7 + 2^7 + 9^7 = 14{,}459{,}929$. This can motivate one to see if an analogous pattern can be constructed. In the coming pages this will be exposed.

We Cannot Neglect the Number 8

Sometimes the beauty of a number evolves from numbers that have repeated digits, as can be seen in Figure 1-1. Notice how the number 8 plays a significant role in the list of products.

1	×	8	=	8
11	×	88	=	968
111	×	888	=	98568
1111	×	8888	=	9874568
11111	×	88888	=	987634568
111111	×	888888	=	98765234568
1111111	×	8888888	=	9876541234568
11111111	×	88888888	=	987654301234568
111111111	×	888888888	=	98765431901234568
111111111	×	888888888	=	987654321791234568
1		8		

Figure 1-1

The Amazing Number 9

As we will notice throughout this book, multiplication is not just a tool of arithmetic. Multiplication can expose the beauties hidden in mathematics. The number 9 provides us with some amazing patterns, such as when we multiply $12{,}345{,}679 \times 9 = 111{,}111{,}111$. If we were to consider multiplying the same number 12,345,679 by the number 9^2, we would expose another unexpected pattern, namely, $12{,}345{,}679 \times 81 = 999{,}999{,}999$. Suppose we multiply the reversal of this number 12,345,679, this time including the number 8 in the sequence. We would arrive at another pattern generated by the number 9, as follows: $987{,}654{,}321 \times 9 = 8{,}888{,}888{,}889$. Notice how the last digit reminds you of what the multiplier was in this calculation.

The calculations in the following Figures 1-2, 1-3, and 1-4 further indicate the power of the number 9.

0	×	9	+	1	=	1
1	×	9	+	2	=	11
12	×	9	+	3	=	111
123	×	9	+	4	=	1,111
1,234	×	9	+	5	=	11,111
12,345	×	9	+	6	=	111,111
123,456	×	9	+	7	=	1,111,111
1,234,567	×	9	+	8	=	11,111,111
12,345,678	×	9	+	9	=	111,111,111

Figure 1-2

The number 9 also provides an opportunity for us to look back at the number 8, which appears repeatedly in Figure 1-3.

$$
\begin{array}{rcccccl}
0 & \times & 9 & + & 8 & = & 8 \\
9 & \times & 9 & + & 7 & = & 88 \\
98 & \times & 9 & + & 6 & = & 888 \\
987 & \times & 9 & + & 5 & = & 8{,}888 \\
9{,}876 & \times & 9 & + & 4 & = & 88{,}888 \\
98{,}765 & \times & 9 & + & 3 & = & 888{,}888 \\
987{,}654 & \times & 9 & + & 2 & = & 8{,}888{,}888 \\
9{,}876{,}543 & \times & 9 & + & 1 & = & 88{,}888{,}888 \\
98{,}765{,}432 & \times & 9 & + & 0 & = & 888{,}888{,}888
\end{array}
$$

Figure 1-3

This time we will use the 9 in another way to generate an amazing pattern, shown in Figure 1-4.

$$
\begin{array}{rclcl}
9 & \times & 9 & = & 81 \\
99 & \times & 99 & = & 9{,}801 \\
999 & \times & 999 & = & 998{,}001 \\
9{,}999 & \times & 9{,}999 & = & 99{,}980{,}001 \\
99{,}999 & \times & 99{,}999 & = & 9{,}999{,}800{,}001 \\
999{,}999 & \times & 999{,}999 & = & 999{,}998{,}000{,}001 \\
9{,}999{,}999 & \times & 9{,}999{,}999 & = & 99{,}999{,}980{,}000{,}001
\end{array}
$$

Figure 1-4

When we consider this list of prime numbers less than 1,000, as shown in Figure 1-5, we notice that the 9 is the only nonprime number (that is, a composite number) found in the units digit of these prime numbers. The other units digits in the list of primes are 1, 2, 3, 5, and 7, all of which are prime numbers.

2, 3, 5, 7, 11, 13, 17, 19, 23, 29, 31, 37, 41, 43, 47, 53, 59, 61, 67, 71, 73, 79, 83, 89, 97, 101, 103, 107, 109, 113, 127, 131, 137, 139, 149, 151, 157, 163, 167, 173, 179, 181, 191, 193, 197, 199, 211, 223, 227, 229, 233, 239, 241, 251, 257, 263, 269, 271, 277, 281, 283, 293, 307, 311, 313, 317, 331, 337, 347, 349, 353, 359, 367, 373, 379, 383, 389, 397, 401, 409, 419, 421, 431, 433, 439, 443, 449, 457, 461, 463, 467, 479, 487, 491, 499, 503, 509, 521, 523, 541, 547, 557, 563, 569, 571, 577, 587, 593, 599, 601, 607, 613, 617, 619, 631, 641, 643, 647, 653, 659, 661, 673, 677, 683, 691, 701, 709, 719, 727, 733, 739, 743, 751, 757, 761, 769, 773, 787, 797, 809, 811, 821, 823, 827, 829, 839, 853, 857, 859, 863, 877, 881, 883, 887, 907, 911, 919, 929, 937, 941, 947, 953, 967, 971, 977, 983, 991, 997

Figure 1-5

It is also fascinating to see how a number can be expressed in a combination of all 10 digits. This is also possible with our number 9 as we can see with the following fractions:

$$9 = \frac{95742}{10683}, \quad 9 = \frac{95823}{10647} \quad 9 = \frac{97524}{10836}.$$

If we place the 0 in the first position of the numbers, we get an additional three fractional equivalents to the number 9, namely,

$$9 = \frac{57429}{06381}, \quad 9 = \frac{58239}{06471}, \quad 9 = \frac{75249}{08361}.$$

The Number System's Key: Number 10

Although the number 10 does not exhibit the same sort of beautiful characteristics as other numbers, we must acknowledge its role in all aspects of our number system, which is the base 10 system. The base 10 system originated in India and was popularized in Europe by Leonardo of Pisa (1170–1240), commonly known as Fibonacci, when he encountered the system in the Arab world. In 1202, Fibonacci published his seminal work *Liber Abaci*, and began by stating he was using "the nine Indian figures: 9, 8, 7, 6, 5, 4, 3, 2, and 1, and with these nine figures and with the sign 0, which the Arabs call *zephyr*, any number whatsoever can be written." Fibonacci encountered these numbers while he was working with mathematicians in Bugia, a town on the Barbary Coast of Africa. It took at least 50 years before these numbers began to be accepted in the European world, where they were known as Hindu-Arabic numerals. That is the history of our base 10 number system. More about Fibonacci in Chapter 5.

The number 10 enables us to generate a prime number in the following strange way:

$$1^1 + 2^2 + 3^3 + 4^4 + 5^5 + 6^6 + 7^7 + 8^8 + 9^9 + 10^{10} = 10{,}405{,}071{,}317,$$

which is a prime number.

The number 10 is also unique in that it is the only number where the sum and the difference of its prime divisors are both prime numbers, since $5 - 2 = 3$, a prime number, and $2 + 5 = 7$, which is also a prime number.

The Number 1 Greater Than 10 Is the Wonderful Number 11

Just as we found strange characteristics of the number 9, which is one less than the number 10, we can also find interesting aspects of the number 11, which is one greater than 10. There are many curiosities embedded in the number 11, such as when we take any number which is a multiple of 11 and reverse the digits. The result will be another number which is also a multiple of 11. For example, the number $135{,}795 = 11 \times 12{,}345$. We reverse the digits of the original number to get $597{,}531 = 54{,}321 \times 11$. We notice that if the original number has 11 as a factor, then the number with the digits reversed also has 11 as a factor.

The number 11 lends itself to some very curious mathematical surprises. Suppose you have a number of several digits, where no two adjacent digits have a sum greater than 9, then you can do the following:

- multiply the number by 11,
- reverse the digits of this product,
- divide the resulting number by 11,

and you will find that you have attained a number whose digits are the reverse of the original number. For example, consider a randomly-selected number 36,181,451 and multiply it by 11 to get $36{,}181{,}451 \times 11 = 397{,}995{,}961$. We now reverse the digits of this product to get 169,599,793, and divide that number by 11 to get

$$\frac{169{,}599{,}793}{11} = 15{,}418{,}163,$$

which quite unexpectedly is a reversal of the original number.

When the number 11 is placed between other multiples of 11, the results are multiples of 11, as we can see with the following numbers:

$$\frac{1111}{11}=101,\quad \frac{2112}{11}=192,\quad \frac{3113}{11}=283,\quad \frac{4114}{11}=374,\quad \frac{5115}{11}=465,$$
$$\frac{6116}{11}=556,\quad \frac{7117}{11}=647.$$

Each of these curiously constructed numbers turns out to be a multiple of 11. This also holds true for larger four-digit products of 11, such as $11\times456=5016$, where we use the same procedure to get

$$\frac{501116}{11}=45556.$$

Therefore, 501,116 is a multiple of 11.

The number 11 also plays a role in the realm of prime numbers, which you recall are numbers that have only two divisors, namely, the number itself and 1.

Here is a cute way of generating an 11-digit number that begins and ends with a number 11. We begin with a repunit number, which is a number consisting of all the same digits. This is done by choosing a number consisting of 11 digits and then subtracting 11-factorial[1] (written 11!):

$$\begin{aligned}&11{,}111{,}111{,}111-11! =\\ &\quad 11{,}111{,}111{,}111-(1\cdot2\cdot3\cdot4\cdot5\cdot6\cdot7\cdot8\cdot9\cdot10\cdot11) =\\ &\quad 11{,}071{,}194{,}311,\end{aligned}$$

which is a prime number.

We can also create an 11-digit prime number by making the first 11 numbers its digits and subtracting 11 as follows:

$$12345678910-11=12{,}345{,}678{,}899,$$

[1] A number factorial (written $n!$) is the product of all the natural numbers to n. That is, $1\cdot2\cdot3\cdot4\cdot5\cdot\cdots n=n!$

which is a prime number. The number 11 can also be used to generate prime numbers in the following ways:

- Even though the following computation is extraordinarily complicated and tiresome, it does show that the number 11 can create a prime number in a very unusual way. That, in itself, can be entertaining. Here is the calculation:

$$\frac{11^{11}\cdot 2^{11}+1}{2^{11}-1}=\frac{584{,}318{,}301{,}411{,}329}{2{,}047}=285{,}451{,}051{,}007,$$

  which is a prime number.
- Furthermore, using all the prime numbers up to the number 11 in the following way generates a prime number: $2 + 3^5 \times 7^{11} =$ 480,490,398,551, which is another prime number.
- Also, the number consisting of 11 digits: 11,111,111,111 = 21649×513239, meaning it is a product of two prime numbers.
- We can also use 11 ones (disguised as other numbers) as follows to generate a prime number: $11^{11}+(11^{11}\times 11)-1=3{,}423{,}740{,}047{,}331$, which is a prime number. Another way in which 11 ones can generate a prime number is as follows: 11!+11!+11!+11!+11!+1 = 199,584,001, which is also a prime number.
- The number 11 is the only prime number where if you add or subtract 6 or 8 from 11, you will always end up with a prime number, as we can see with the following:

$$11 + 6 = 17,\ 11 - 6 = 5,\ 11 + 8 = 19, \text{ and } 11 - 8 = 3.$$

- The number 11 is the smallest prime number that can be expressed as the sum and product of 2 other prime numbers, such as: $(2+3) + (2\times 3) = 11$.
- The number 11 squared can be expressed as a sum of 3 consecutive prime numbers:

$$11^2 = 121 = 37 + 41 + 43.$$

- The number 11 cubed can be expressed as a sum of 3 consecutive prime numbers as well:

$$11^3 = 1331 = 439 + 443 + 449.$$

- We get a surprise package when we compute the sum of the 11 primes after the number 11, namely, $13 + 17 + 19 + 23 + 29 + 31 + 37 + 41 + 43 + 47 + 53 = 353$, which is a palindromic number (see page 92), and the 71st prime number.

The curiosities with the number 11 are offered to amuse the reader further about this amazing number. As a final amusement, consider the number 11^2, which can be expressed as the sum of five consecutive powers of 3 as follows: $11^2 = 121 = 3^0 + 3^1 + 3^2 + 3^3 + 3^4$. If that is not impressive enough, we also have 11^3 as the sum of the squares of three consecutive odd numbers: $11^3 = 1331 = 19^2 + 21^2 + 23^2$. These are just some of the amazing characteristics of the number 11 that hopefully will motivate the reader to pursue other curiosities of this number.

The Curious Number 12

The number 12 has a peculiarity that is shared by very few other numbers. Consider $12^2 = 144$, and the square of the number with its digits reversed: $21^2 = 441$. The results are reversals of one another.

The number 12 seems to appear in unusual and unexpected circumstances. For example, if we take the square of any prime number greater than 3 and subtract 1 from it, we get a multiple of 12. To convince you that this is true, we will take a few randomly selected prime numbers and do the previously described calculation:

$$5^2 - 1 = 24 = 12 \times 2, \quad 13^2 - 1 = 168 = 12 \times 14, \quad 23^2 - 1 = 528 = 12 \times 44,$$
$$37^2 - 1 = 1368 = 12 \times 114.$$

We notice that each of these calculations results in a product involving number 12.

There are also coincidences in mathematics that are just a bit difficult to justify, but easily appreciated. One such involves the number 12. Here is a true coincidence, where the 12th prime, 37, is a reversal of the 21st prime, 73. Notice both reversals!

The Not-So-Unlucky Number 13

An analogous situation to the last one with the number 12 can be repeated with the number 13 as follows: $13^2 = 169$ and $31^2 = 961$, so that the reversal of the number 13 squared is the square of 31, or 13 reversed. It is also curious that the prime number 13 is one of those numbers whose reversal is also a prime number, namely, 31. Furthermore, the number 13 is the smallest number that can be expressed as the sum of the squares of two prime numbers, namely, $2^2 + 3^2 = 13$.

There are a host of fascinating numerical relationships involving the number 13, some of which we provide here:

- The sum of the first 6 prime numbers, $2+3+5+7+11+13 = 41$, just happens to be the 13th prime number.
- If we divide 13 by each of the first prime numbers less than 13, the sum of the remainders $1 + 1 + 3 + 6 + 2 = 13$.
- The number 13 is the smallest prime number where the sum of the digits $1 + 3 = 4$ is a perfect square.
- The sum of the first 13 prime numbers $2 + 3 + 5 + 7 + 11 + 13 + 17 + 19 + 23 + 29 + 31 + 37 + 41 = 238$, whose digit sum is 13.
- A prime number can be generated by applying the number 13 in a rather strange way. The numerator of the following fraction has 14 digits, of which 13 digits are 3s:

  $$\frac{13{,}333{,}333{,}333{,}333}{13} = 1{,}025{,}641{,}025{,}641,$$

  which is a prime number. Not only that, but a highly unusual and symmetric number where triples of digits repeat.
- There are exactly 13 prime numbers consisting of four consecutive digits, and they are: 1423, 2143, 2341, 2543, 4231, 4253, 4523, 4567, 4657, 5647, 5867, 6547, 6857.
- The number of composite numbers between 13 and its reversal, 31, is 13. These composite numbers are: 14, 15, 16, 18, 20, 21, 22, 24, 25, 26, 27, 28, and 30.
- We can represent the number 13 in terms of all the prime numbers less than 13 as follows: $(5 \times 11) - (2 \times 3 \times 7) = 13$. As a challenge, one could seek other ways that prime numbers in combination can create number 13.

The Unique Number 16

The number 16 is the only square integer that can be expressed as $x^y = y^x$, in that $2^4 = 4^2$. This uniqueness was proved by the famous Swiss mathematician Leonhard Euler (1707–1783).

There is an interesting pattern for generating prime numbers, where the number 16 is the largest number that allows this to happen. We first add 16 to itself and subtract 1, then add half of 16 to itself and subtract 1, and so on. Each of these will result in a prime number as follows:

$$16 + 16 - 1 = \mathbf{31},\ 16 + 8 - 1 = \mathbf{23},\ 16 + 4 - 1 = \mathbf{19},\ 16 + 2 - 1 = \mathbf{17}.$$

Each of these is a prime number.

As a tribute to its uniqueness, the number 16 is the largest known integer, n, where the number $2^n + 1$ is a prime number. In other words, $2^{16} + 1 = 65{,}536 + 1 = 65{,}537$, which is the 6,543rd prime number. No greater value of n would produce a larger prime number.

There are times when we search in unusual ways for the presence of a number, as is the case here with the number 16. The sum of the 16th prime number (53) and the 16th composite number (26) is a prime number, $53 + 26 = 79$. Furthermore, the sum of those digits is 16, that is, $7 + 9 = 16$.

The number 16 also appears in geometry when considering a square whose side length is 4, which then has an area numerically equal to its perimeter, 16.

Surprises Presented Through the Number 17

In mathematics, the number 17 possesses some unusual properties. Not only is the number 17 a prime number, but it is also the sum of the first four prime numbers, namely, $2 + 3 + 5 + 7 = 17$. It should be noted that this is the only time that the sum of four consecutive prime numbers could result in another prime, since any of the other four consecutive primes would not include an even number, and therefore, would result in a sum of four odd numbers, which is an even number, and therefore, not a prime number, since the only even prime number is the number 2.

There are quite a few mind-boggling properties of the number 17. We offer just some of these here as entertainment:

- The number 17 is the smallest prime number the sum of whose digits is a perfect cube, since $1 + 7 = 8 = 2^3$.
- The number 17 can also be expressed as the sum of two fourth powers, making it the smallest prime with this feature. That is, $1^4 + 2^4 = 1 + 16 = 17$.
- Moreover, the number 17 is also the smallest prime number that can be expressed as the sum of the squared prime and a cubed prime in two different ways, namely, $3^2 + 2^3 = 9 + 8 = 17$ and $1^3 + 4^2 = 1 + 16 = 17$. This is the only prime number where consecutive primes can be used in this way.
- The number 17 is a smallest prime that is the average of seven consecutive prime numbers, namely, 7, 11, 13, 17, 19, 23, 29. We get the average as

$$\frac{7+11+13+17+19+23+29}{7} = \frac{119}{7} = 17.$$

  We should also note that the number 17 is the smallest prime that is the average of two squared primes, namely,

$$\frac{3^2+5^2}{2} = \frac{9+25}{2} = \frac{34}{2} = 17.$$

- The number 17 is also the first member of a "twin prime" (twin primes are prime numbers whose difference is 2) pair with the number 19. If you place a 9 in the middle of these two twin primes, you get 197 and 199, another pair of twin primes.
- Sometimes one has to search for unusual appearances of the number 17. This is the case with $2^{17} = 131{,}072$, which is not a prime number. However, its reversal, 270,131, is indeed the 23,652nd prime number.
- Here is a curiosity that is quite astonishing: $7{,}413{,}245{,}658 = 17^7 + 17^4 + 17^1 + 17^3 + 17^2 + 17^4 + 17^5 + 17^6 + 17^5 + 17^8$, which at first glance is not notable. But if one looks at the sequence of exponents and compares it to the original number, the result is truly bewildering.

- The number 17 can also generate 11 prime numbers with the expression $n^3 + n^2 + 17$, when n takes on the values from 0 to 10.
- The number 17 belongs to a group of numbers that share some strange characteristics. For example, $17^3 = 4913$, and the sum of the digits of this number is $4 + 9 + 1 + 3 = 17$. The only other numbers that share this characteristic are 1, 8, 18, 26, and 27. You might want to check this to be convinced of this property. We offer one example here: $26^3 = 17{,}576$, and $1 + 7 + 5 + 7 + 6 = 26$.

The Lucky Number 18

We can also have some fun with the lucky number 18. Begin by taking any three-digit number whose digits are all different, and then arrange them in order to form the largest number, and then the smallest number. Subtract the smaller number from the larger number, and you will find that this answer will be a number the sum of whose digits is 18. Let's consider an example. We shall select the number 584. We then write the smallest number with these digits (458) and then the largest number with these digits (854). Subtracting these numbers 854 – 458 = 396. The sum of the digits of this number (3+9+6) is 18. Try it with some other three-digit number to convince yourself of this property.

By inserting a 9 between the two digits of 18, we get 198 = 99 + 99. And again reversing the digits, we get 891 = 9 × 99. Also notice that 18 + 81 = 99, and 9 + 9 = 18. We can extend this strange relationship by inserting more and more 9s between the numbers 18 and 81, as shown in Figure 1-6. Note that this pattern continues further.

18	=	9 + 9		81	=	9 × 9
198	=	99 + 99		891	=	9 × 99
1998	=	999 + 999		8991	=	9 × 999
19998	=	9999 + 9999		89991	=	9 × 9999

Figure 1-6

The number 18 does not harbor too many mathematical curiosities. One such is that $18 = 9 + 9$, and 9×9 is equal to its reversal, 81. As we search for more peculiarities, we can notice that 18 is the smallest number that is twice the sum of its digits. One can generate two operations with the number 18 that will yield two numbers that use all 10 digits exactly once as follows: 18^3 = **5,832**, and 18^4 = **104,976**.

When we consider the 3rd power of 18, the 6th power of 18, and the 7th power of 18, we find that each provides a number whose digits have a sum of 18, as shown below:

$$18^3 = 5{,}832 \text{ and } 5 + 8 + 3 + 2 = 18,$$

$$18^6 = 34{,}012{,}224 \text{ and } 3 + 4 + 0 + 1 + 2 + 2 + 2 + 4 = 18,$$

$$18^7 = 612{,}220{,}032 \text{ and } 6 + 1 + 2 + 2 + 2 + 0 + 0 + 3 + 2 = 18.$$

In geometry, the number 18 is one of two numbers where a rectangle has a numerical measure of its area equal to the numerical measure its perimeter. This rectangle is one where the length and width are 6 and 3, respectively, and where the area is 3×6 and the perimeter is $3 + 3 + 6 + 6$, each of which are numerically equal to 18.

Sparse Properties of the Number 19

Similarly to 17, we have a situation where a number's digits can be made exponents of the number 19, and the sum of these powers is equal to that original number. This is the case when the digits of the number 52,135,640 are the exponents of the number 19, and the sum of these powers $19^5 + 19^2 + 19^1 + 19^3 + 19^5 + 19^6 + 19^4 + 19^0 =$ 52,135,640. Notice the relationship between the sequence of exponents and the digits of the sum.

Another curiosity involving the number 19 is that the sum of the first 19 numbers is 10 times the number 19, as shown here: $19 + 18 + 17 + 16 + 15 + 14 + 13 + 12 + 11 + 10 + 9 + 8 + 7 + 6 + 5 + 4 + 3 + 2 + 1 + 0 = 190$.

Here are ways the prime number 19 relates to other primes:

- If one takes the sum of the digits of the number 19 and places this sum between 19's two digits to get the number 1109, a prime number results.
- The number 19 is the smallest prime number that is equal to the sum of the product of its digits and the sum of its digits, as we see: $19 = (1 \times 9) + (1 + 9)$.
- The number 19 is the smallest prime number that contains a digit that is not a prime number.
- The number 19 is the smallest prime number whose reversal, 91, is not a prime number.
- The number 19 is the only prime number that is equal to the difference of two cubed primes, namely, $3^3 - 2^3 = 19$.
- The number 19 is the smallest prime where the sum of the digits of 19 and $19^2 = 361$ are each equal to 10.
- Here we show how we can use 19 ones to create a prime number:
- $1{,}111 \times 111^{11} + 111{,}111{,}111 + 1 = 35{,}016{,}023{,}542{,}301{,}708{,}082{,}592{,}033$, which is a prime number.
- There are times when a nifty pattern up to the number 19 can unexpectedly generate a prime number. In a sequence where the bases are the initial prime numbers up to 19 and the exponents are the consecutive natural numbers, we get the sum

  $$2^1 + 3^2 + 5^3 + 7^4 + 11^5 + 13^6 + 17^7 + 19^8 = 17{,}398{,}892{,}111,$$

  which is a prime number.

One often neglected curiosity is that the digit 8 appears 19 times in the numbers from 1 to 100.

Appearances of the Number 23

Consider a few cute arrangements with the digits of 23:

- An unexpected symmetry for the number 23 is the operation where the summands increase from 0 to 5 and their exponents decrease from 5 to 0, thus producing the number 23, as shown here: $0^5 + 1^4 + 2^3 + 3^2 + 4^1 + 5^0 = 0 + 1 + 8 + 9 + 4 + 1 = 23$.

- Another sequential relationship that will generate the number 23 involves taking the first three prime numbers, 2, 3, and 5, and then multiplying them consecutively in the following way:

$$(2) + (2 \times 3) + (3 \times 5) = 2 + 6 + 15 = 23.$$

- We also have coincidences in mathematics concerning the number 23. The sum of the first 23 prime numbers, $2 + 3 + 5 + 7 + 11 + 13 + 17 + 19 + 23 + 29 + 31 + 37 + 41 + 43 + 47 + 53 + 59 + 61 + 67 + 71 + 73 + 79 + 83 = 874$, which is a product of 23, namely, $874 = 23 \times 38$. Notice again that the 23rd prime happens to be 83, which is reversal of the other factor of the product.
- Suppose we find the sum of the digits of the number 23^{2+3}, which is 6,436,343. Now take the digits' sum to get

$$6 + 4 + 3 + 6 + 3 + 4 + 3 = 23.$$

  What a surprise!
- We shouldn't be limited to the examples shown here because there are times when one can choose two numbers, which are practically unrelated, and create the desired number, in this case 23. Such is the situation when we use the numbers 3 and 5 to get the following:

$$\frac{3^5+5^3}{(3+5)+(5+3)} = \frac{368}{16} = 23.$$

The sum of the first three primes $2 + 3 + 5 = 10$. The next power of 10, which is 100, can be achieved by taking the consecutive primes up to the number 23, such as: $2 + 3 + 5 + 7 + 11 + 13 + 17 + 19 + 23 = 100$. It is not known whether any other consecutive listing of primes will lead to a higher power of 10.

Simple Characteristics of the Number 24

In mathematics, the number 24 harbors simple peculiarities, such as where 24 is divisible by both the sum of its digits (6) and the product of its digits (8). Also, the number 24 is the smallest number that has exactly eight divisors, namely, 1, 2, 3, 4, 6, 8, 12, and 24. If we now subtract 1 from each of these divisors (except the first two) we will end up with six prime numbers. Furthermore, if we take the product of the proper divisors of 24, a perfect cube results, namely, $2 \times 3 \times 4 \times 6 \times 8 \times 12 = 13{,}824 = 24^3$. Curiously enough, if we take the sum of the squares of the first 24 numbers, we end up with another square number, as we can see here:

$$1^2 + 2^2 + 3^2 + 4^2 \cdots 23^2 + 24^2 = 4900 = 70^2.$$

As we move along, we see that the number 24 has a unique relationship with the number 26. We take the sum of the squares of the divisors of 24: $1^2 + 2^2 + 3^2 + 4^2 + 6^2 + 8^2 + 12^2 + 24^2 = 850$, and now take the sum of the squares of the divisors of 26: $1^2 + 2^2 + 13^2 + 26^2 = 850$. This is a good lead-in to the next number: 26.

A Well-Hidden Surprise with the Number 26

The number 26 does not enjoy many curiosities, especially since it is placed between a perfect square ($5^2 = 25$) and a perfect cube ($9^3 = 27$). We would have to go to $26^2 = 676$ to get something special, in this case, a palindromic number. Yet if we look at $26^3 = 17{,}576$, and then take the sum of these digits: 1 + 7 + 5 + 7 + 6, we get 26.

A curious aspect occurs when we take 10^{26} and add or subtract the number 123456789.

In the first case, $123456789 + 10^{26}$, we get the number 100,000, 000,000,000,000,123,456,789, and in the second case, $10^{26} -$ 123456789, we get 99,999,999,999,999,999,876,543,211. As attractive as these numbers might be, the amazing thing is that both of these numbers are prime numbers. We find something truly amazing when we notice what the sum of these two numbers is 200,000,000,000,000,000,000,000,000.

Some Simple Relationships Involving the Number 27

Number 27 has a rather unique place amongst two-digit numbers (which it shares with the number 37). This is that every three-digit multiple of 27, such as 27 × 13 = 351, when *cyclically* altered (for example, the various cyclical alternatives to 351 are 513, and 135), generates another multiple of 27 (19 × 27 = 351, and 5 × 27 = 135).

To further clarify this curious characteristic, consider another multiple of 27, such as $27 \times 17 = 459$. The cyclical alternatives are 594 and 945, each of which is a multiple of 27, as $594 = 27 \times 22$, and $945 = 27 \times 35$.

A cute pattern for the number 27 can be created if you add the consecutive numbers between the digits 2 and 7, you will get $2 + 3 + 4 + 5 + 6 + 7 = 27$. Another curiosity to consider is that if we take the sum of the digits of $27^3 = 19{,}683$, we get $1 + 9 + 6 + 8 + 3 = 27$. Furthermore, the number 27 is the smallest number that can be expressed as a sum of three squares in two different ways: $3^2 + 3^2 + 3^2 = 27 = 5^2 + 1^2 + 1^2$.

The Perfect Number 28

The number 28 there possesses two very distinct characteristics. First, it is the second *perfect number* (after the perfect number 6), since the sum of its factors, $1 + 2 + 4 + 7 + 14$, is equal to 28. The next few perfect numbers after the number 28 are:

496
8128
33,550,336
8,589,869,056
137,438,691,328

Notice how the perfect numbers all have a units digit of 6 or 8. If we add the number 28 to the sum of its factors to get the sum $1 + 2 + 4 + 7 + 14 + 28$, and then take the reciprocals of these numbers, we get

$$\frac{1}{1} + \frac{1}{2} + \frac{1}{4} + \frac{1}{7} + \frac{1}{14} + \frac{1}{28} = 2,$$

which is true for all perfect numbers.

From an arithmetical perspective, the number 28 has a very unusual characteristic in that it can be expressed as a sum of the first positive integers: $1 + 2 + 3 + 4 + 5 + 6 + 7 = 28$; it can be expressed as a sum of the first prime numbers: $2 + 3 + 5 + 7 + 11 = 28$; and it can also be expressed as the sum of the first non-prime numbers: $1 + 4 + 6 + 8 + 9 = 28$. Let's not forget that 28 can also be expressed as sum of two cubes, as $28 = 1^3 + 3^3$.

The Relatively Small Prime Number 29

The number 29 is the smallest two-digit prime number equal to the sum of the digits of all the previous two-digit primes. In other words, the sum of the digits of 11, 13, 17, 19, and 23, or $1 + 1 + 1 + 3 + 1 + 7 + 1 + 9 + 2 + 3 = 29$.

The number 29 is the smallest multi-digit prime number where the product of the digits of the cube of the number is also perfect cube. We can see that $29^3 = 24{,}389$, and the product of the digits of the number 24,389 is also a perfect cube, as $2 \times 4 \times 3 \times 8 \times 9 = 1{,}728 = 12^3$.

There is also a special uniqueness to the number 29, as it is the only two-digit prime number whose square is the sum of the squares of two consecutive two-digit numbers. That is, $20^2 + 21^2 = 400 + 441 = 841 = 29^2$.

The square of the number 29 is the last number in a series of consecutive squares whose sum is a square number and whose square root happens to be the reverse of 29, as shown here:

$$7^2 + 8^2 + 9^2 + \cdots + 27^2 + 28^2 + 29^2 = 8{,}464 = 92^2.$$

As we search for more curiosities involving the number 29, we find that $2^2 + 3^2 + 4^2 = 29$, making 29 the smallest prime number equal to the sum of three consecutive squares. And further, the number 29 is the smallest prime that can be expressed in terms of four consecutive prime numbers as follows: $(2 \times 7) + (3 \times 5)$.

A Magical Surprise with the Number 30

The number 30 is the smallest number which has 3 distinct prime factors, namely, 2, 3, and 5. If we take the reciprocals of these factors and subtract the reciprocal of 30, we get the following surprise:

$$\frac{1}{2} + \frac{1}{3} + \frac{1}{5} - \frac{1}{30} = 1.$$

The number 30 can be shown to be the sum of the first four square numbers, as we see here:

$$1^2 + 2^2 + 3^2 + 4^2 = 1 + 4 + 9 + 16 = 30.$$

The Tricky Number 31

A fun aspect of the number 31 is found by squaring the number 31 to get $31^2 = 961$, and then rotating it 180° to get another perfect square: $196 = 14^2$.

The prime numbers leading up to number 31 are: 2, 3, 5, 7, 11, 13, 17, 19, 23, 29, 31. When we take the sum of the digits of these prime numbers, we get:

$$2+3+5+7+1+1+1+3+1+7+1+9+2+3+2+9+3+1=61,$$

which is a prime number. We also should note that $31^7 = 27{,}512{,}614{,}111$, where the sum of the digits is $2+7+5+1+2+6+1+4+1+1+1=31$.

We must be careful not to generalize, as we see in the following situation. The number 31 is the lead number for a sequence of curious looking numbers that are all primes: 31, 331, 3331, 33331, 333331, 3333331, 33333331. However, the next numbers in the sequence are no longer prime numbers, since $333333331 = 17 \times 19{,}607{,}843$, and $3333333331 = 673 \times 4{,}952{,}947$, and so on. Therefore, caution must be exercised before one generalizes a situation that appears to support a pattern but in actuality doesn't.

The number 31 can also be expressed in some interesting ways. For example, $31 = 2^0 + 2^1 + 2^2 + 2^3 + 2^4$, and $31 = 5^0 + 5^1 + 5^2$. The only number other than 31 with 1 as the units digit that can be expressed as a sum of successive powers is the number 8191.

We have *happy numbers* in mathematics, which as we mentioned earlier, are numbers where summing the squares of the digits and repeating this process will lead to the number 1. The number 31 is a happy number, as $3^2 + 1^2 = 10$ and $1^2 + 0^2 = 1$. The number 31 is also the first of a pair of happy numbers with 32, since $3^2 + 2^2 = 13$, and $1^2 + 3^2 = 10$ and $1^2 + 0^2 = 1$. (See page 100 for more on happy numbers.)

Here is one that is truly amazing. Consider the number 31,111, 111,111,111,111,111,111,111,111,111, which has 31 unit-digits. If we divide that number by 31, we get 1,003,584,229,390,681,003, 584,229,390,681, which is a prime number.

The trickiness of the number 31 can be quite surprising. Consider the following sequence and find the next number: **1, 2, 4, 8, 16.** Clearly most people would guess that 32 is the next number. Yes, that would be fine. However, when the next number, surprisingly, is given as 31 (instead of the expected 32), much to our amazement this is a correct answer as well, and **1, 2, 4, 8, 16, 31** is a legitimate sequence.

We will justify this strange appearance a bit later, but in the meantime, let us first find the succeeding numbers in this unexpected sequence. We shall set up a chart showing the differences between the terms of the given sequence up to 31, then find the difference between these differences, and then work backwards to establish a pattern. Here, we can establish a pattern at the third difference, as shown in Figure 1-7.

Original Sequence	**1**		**2**		**4**		**8**		**16**		**31**
First Difference		1		2		4		8		15	
Second Difference			1		2		4		7		
Third Difference				1		2		3			
Fourth Difference					**1**		**1**				

Figure 1-7

With the fourth differences forming a sequence of constants, we can reverse the process (turn the table upside down) and extend the third differences a few more steps with 4 and 5, to find the next two numbers in the sequence after 31: the numbers 57 and 99, as shown in Figure 1-8.

Fourth Difference					1		1		1		1				
Third Difference				1		2		3		**4**		**5**			
Second Difference			1		2		4		7		**11**		**16**		
First Difference		1		2		4		8		**15**		**26**		**42**	
Original Sequence	1		2		4		8		16		**31**		**57**		**99**

Figure 1-8

The bold-type numbers are those that were obtained by working backwards from the third difference sequence. The general term is a fourth-power expression since we had to go to the third differences to get a constant. The general term (n) is:

$$\frac{n^4 - 6n^3 + 23n^2 - 18n + 24}{24}.$$

A geometric interpretation should help convince you of the beauty and consistency inherent in mathematics. Let's count the number of regions into which a circle can be partitioned by joining points on the circle. We make sure that no three lines meet at one point, or else you will lose a region. When we do this with 6 points, as shown in Figure 1-9, we find that there are 31 regions. The chart in Figure 1-10 summarizes the number of regions based on specific points on the circle.

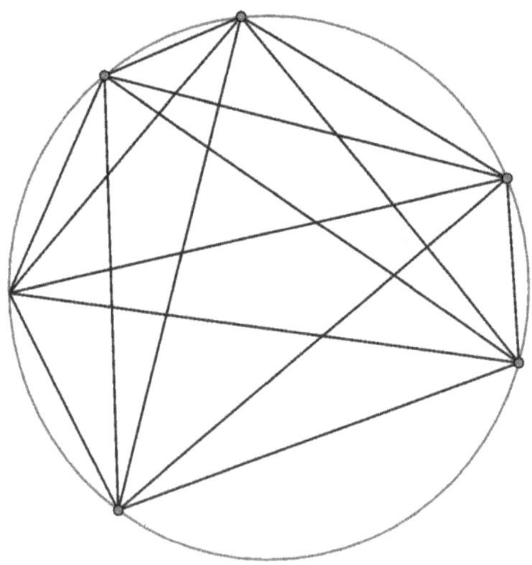

Figure 1-9

Number of points on the circle	Number of regions into which the circle is partitioned
1	1
2	2
3	4
4	8
5	16
6	**31**
7	57
8	99

Figure 1-10

The Secret Number 33

It is believed that there are no sequences of consecutive natural numbers including *no* prime numbers longer than the sequence of 33 numbers shown here:

1328, 1329, 1330, 1331, 1332, 1333, 1334, 1335, 1336, 1337, 1338, 1339, 1340, 1341, 1342, 1343, 1344.
1345, 1346, 1347, 1348, 1349, 1350, 1351, 1352, 1353, 1354, 1355, 1356, 1357, 1358, 1359, 1360.

There are some curious ways that the number 33 can generate prime numbers, such as the following:

$$33 \times 4^{33} + 1 = 2{,}434{,}970{,}217{,}729{,}660{,}813{,}313,$$

$$33 \times 14^{33} + 1 = 2{,}33191{,}488{,}102{,}653{,}841{,}084{,}647{,}604{,}804{,}502{,}947{,}889{,}153,$$

$$33 \times 34^{33} + 1 = 11{,}410{,}855{,}408{,}151{,}892{,}011{,}078{,}343{,}133{,}624{,}756{,}861{,}835{,}061{,}265{,}170{,}433.$$

But $33 \times \mathbf{24}^{33} + 1 = 116{,}274{,}616{,}654{,}336{,}630{,}549{,}169{,}399{,}022{,}016{,}368{,}650{,}005{,}512{,}193$, which is *not* a prime number.

The Smallest Number 34

The number 34 is the smallest number that can be expressed as a sum of 2 primes in 4 different ways, as follows: $34 = 3 + 31 = 5 + 29 = 11 + 23 = 17 + 17$. Furthermore, the number 34 is the smallest number that can be expressed as a sum of distinct primes, each of which is comprised only of odd digits, as follows: $34 = 3 + 5 + 7 + 19$. Notice, all the single-digit odd numbers have been used exactly once.

The Challenge for the Number 35

The number 35 can be written as a sum of odd primes in 35 different ways. For example: $35 = 29 + 3 + 3 = 23 + 3 + 3 + 3 + 3 = 11 + 7 + 3 + 3 + 3 + 3 + 5$. The motivated reader may wish to find the remaining 32 ways to express the number 35 as the sum of odd primes numbers.

More Than Just a Square: The Number 36

The number 36 is well-known as a square number, since $6^2 = 36$. The number 36 is also the smallest square number that can be expressed as a sum of a pair of twin prime numbers, namely, $36 = 17 + 19$. Furthermore, beyond the sum of twin primes, the number 36 is the smallest number which can be expressed as a sum of two distinct primes in three different ways, as follows: $36 = 5 + 31 = 7 + 29 = 13 + 23$. We could also extend this by noting that the number 36 can be expressed in two different ways as a sum of consecutive prime numbers: $36 = 5 + 7 + 11 + 13$, or $36 = 17 + 19$.

Partnering with its reversal, the number 36 is the smallest multi-digit number that generates prime numbers in the following fashion: $36 \times 63 - 1 = 2267$, and $36 \times 63 + 1 = 2269$, producing a pair of twin primes, that is, the 336th prime and the 337th prime, respectively.

The number 36 is also the eighth triangular number, as we can see in Figure 1-11. This of course can be shown arithmetically as follows: $1 + 2 + 3 + 4 + 5 + 6 + 7 + 8 = 36$. (See page 94.)

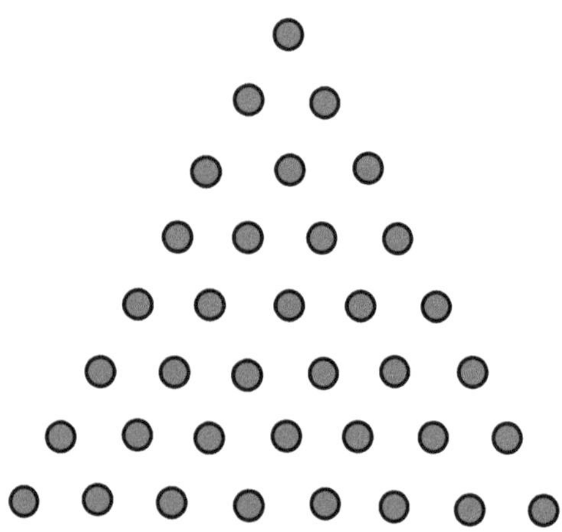

Figure 1-11

The number 36 also has the distinction of being the largest 2-digit number that is divisible by both the product of its digits (3×6) and the sum of its digits (3+6). Somewhat analogously, we can show a neat product and sum that can equal the number 36:

$$1^2 \times 2^2 \times 3^2 = 1 \times 4 \times 9 = 36, \text{ and } 1^3 + 2^3 + 3^3 = 1 + 8 + 27 = 36.$$

The Pattern-Generating Number 37

We begin our discussion of the number 37 with the following pattern:

$$\begin{aligned}
37 \times (1+1+1) &= 111\\
37 \times (2+2+2) &= 222\\
37 \times (3+3+3) &= 333\\
37 \times (4+4+4) &= 444\\
37 \times (5+5+5) &= 555\\
37 \times (6+6+6) &= 666\\
37 \times (7+7+7) &= 777\\
37 \times (8+8+8) &= 888\\
37 \times (9+9+9) &= 999
\end{aligned}$$

Not only is the number 37 a prime number, but its reversal, 73, is also a prime number. Furthermore, each of these prime numbers can be expressed as a sum of two squares: $37 = 1^2 + 6^2$, and $73 = 3^2 + 8^2$. The number 37 can also be found by taking the sum of the first five consecutive composite numbers: $37 = 4 + 6 + 8 + 9 + 10$. We can also create 37 as a sum using only the number 3 as follows:

$$37 = 33 + 3 + \frac{3}{3}.$$

We should also note that the number 37 is the only prime number whose reciprocal has a period of length three, as we see as follows:

$$\frac{1}{37} = 0.027\ 027\ 027\ 027\ 027\ldots.$$

The number 37 has some peculiar characteristics. For example, suppose we take a three-digit number which is divisible by 37, such as $296 = 8 \times 37$. We will find that 962 and 629 are two numbers where the digits have been cyclically permutated (same order, but different starting point) from the original number 296. Each of these two numbers is also divisible by 37, since $962 = 26 \times 37$, and $629 = 17 \times 37$.

The number 37 lends itself to another rather amazing property. If you take any multiple of 37, break it up into two numbers, and then find the sum of the cubes of these two parts, the result will always be a multiple of 37. Let's consider a few examples: $37 \times 5 = 185$. Suppose we split 185 into 18 and 5 and then add the cubes of each part: $18^3 + 5^3 = 5957 = 37 \times 161$. We could have split the number 185 as 1 and 85 and take the sum of the cubes of these 2 numbers, that is, $1^3 + 25^3 = 614,125 + 1 = 614,126 = 16598 \times 37$. This time take the number $37 \times 8 = 296$ and split the number as 2 and 96. The sum of the cubes is $2^3 + 96^3 = 8 + 884,736 = 884,744 = 37 \times 23,912$, once again a multiple of 37. We could have also split the number 296 as 29 and 6, where we get the sum of the cubes $29^3 + 6^3 = 24,605 = 37 \times 665$. An ambitious reader might want to try this with other multiples of 37.

The Primes Related to Composite Number 38

We know that in mathematical terms the number 38 is not a prime number. Yet, we can show that 38 is related to prime numbers in that it is the sum of the squares of the first three prime numbers, namely, $2^2 + 3^2 + 5^2 = 38$. Furthermore, we can also create a prime number from the digits of 38, for example, $3 + 8 = 11$. We can also get a prime number in a more complicated way as follows: $3^2 + 8^{19} = 9 + 144,115,188,075,855,872 = 144,115,188,075,855,881$, which is a prime number.

The number 38 is the largest even number that can be expressed in only one way as the sum of two distinct prime numbers, as $38 = 31 + 7$. It is also the largest even number that can be expressed as the sum of two odd numbers, where at least one of the numbers is a prime number. An ambitious reader may want to check and verify that 38 is the largest even number that is not the sum of two odd composite numbers.

The number 38 is a composite number that determines a very unusual characteristic, which is, if we add 1s onto the number 38, we will always have another composite number, such as the following numbers: 381, 3811, 38,111, 381,111, and 3,811,111.

Expressing the Number 39 in Various Ways

The number 39 can be expressed as the sum of the product and the sum of its digits as we show with the following: $(3 \times 9) + (3 + 9) = 39$. The number 39 can also be expressed as a sum of consecutive prime numbers: $3 + 5 + 7 + 11 + 13 = 39$, and also as a product of the first and last of those prime numbers: $3 \times 13 = 39$. Moreover, the number 39 can also be expressed as the sum of the first 3 powers of 3 as follows: $3^1 + 3^2 + 3^3 = 39$.

Tricks with the Number 41

The number 41 lends itself to an amazing mathematical experience. This technique, which has a really surprising ending, can be created by beginning with the number 41, then adding 2 to get 41 + 2 = **43**, then adding 4 to that number to get 43 + 4 = **47**, then adding 6 to get **53**, then adding 8 to get **61**, then adding 10 to get **71**, then adding 12 to get **83**. The next number in this scheme would be **97**. By continuing this process 40 times you will have a string of 40 prime numbers. However, when you do it for the 41st time you will end up with a composite number!

Here is a peculiar challenge that appears to be true, but we leave it to the motivated reader to test out several possibilities. Select a pair of numbers whose sum is 41. Square one of the numbers and add the other number to the squared number, and you should end up with a prime number. For example, suppose we select the numbers 33 and 8, whose sum is 41. Following this procedure: $33^2 + 8 = 1097$, which is a prime number. Or we can do it in the other direction: $8^2 + 33 = 97$, which is also a prime number.

A quick trick method to multiply by 41 is as follows: quadruple the number you wish to multiply by 41 and then multiply the result by 10 and add the original number. To see how this works, consider the following example. To multiply 47 by 41, quadruple 47, which yields 188, then multiply by 10 to get 1880, and then add the original number 47 to get 1927, which is equal to 47×41.

The number 41 is also the smallest prime number the sum of whose digits, 5, and the difference of the digits, 3, form a pair of twin primes (3 and 5), which, as we recall, are two prime numbers whose difference is 2.

The number 41 can be expressed as the sum of the first 6 prime numbers, as $2 + 3 + 5 + 7 + 11 + 13 = 41$, and also can be expressed, simply, as a sum of 3 other prime numbers, as $11 + 13 + 17 = 41$.

Generating Prime Numbers with the Number 42

Patterns can create an interesting group of five prime numbers using the number 42:

$1 + 42 = 43$, which is the 14th prime number
$1 + 4242 = 4243$, which is the 582nd prime number
$1 + 424242 = 424243$, which is the 35,721st prime number
$1 + 42424242 = 42424243$, which is the 2,571,845th prime number
$1 + 4242424242 = 4242424243$, which is also a prime number!

The number 42 is a composite number and yet the product of 3 distinct prime numbers, namely, $42 = 2 \times 3 \times 7$. The number 42 is also the sum of the first 6 positive even integers, as $42 = 2 + 4 + 6 + 8 + 10 + 12$. Furthermore, the number 42 is also equal to the sum of its non-prime proper divisors: $42 = 1 + 6 + 14 + 21$.

Now the next feature will surely amaze you, however, it is not doable without a calculator/computer. In 2019, mathematicians found that the number 42 can be expressed as the sum of three cubes as follows:

$$(-80{,}538{,}738{,}812{,}075{,}974)^3 + (80{,}435{,}758{,}145{,}817{,}515)^3 + (12{,}602{,}123{,}297{,}335{,}631)^3.$$

The number 42 also lends itself to a curious relationship between the cubes of two prime twins, namely, $42^2 = 11^3 + 433 = 13^3 - 433$.

The Cool Number 43

There are always surprising coincidences in mathematics, such as when the number 43 is raised to the $(4 + 3 = 7)$ power. The result is a number that has a digit sum of 43, as we can see with the following: $43^7 =$ 271,818,611,107, where $2 + 7 + 1 + 8 + 1 + 8 + 6 + 1 + 1 + 1 + 0 + 7 = 43$. A group of five primes are formed by a pattern within the number 43, as shown in Figure 1-12.

43
4483
444883
44448883
4444488883

Figure 1-12

Mathematicians often discover personal curiosities with numbers. For example, the English mathematician Augustus De Morgan (1806–1871) was a man of wisecracks, which he felt were humorous. There were times when he found mathematical coincidences, such as with the number 43. He claimed to have the good fortune of being x years old in the year x^2. That is, he was 43 years old in 1849, and surprisingly, $43^2 = 1849$.

The number 43 is the smallest non-palindromic prime that results in a square number when its reversal is subtracted from it, as we can see with $43 - 34 = 9 = 3^2$. Another peculiarity to consider is that the number 43 is the only prime number where using its digits sequentially as both base and power yields $4^4 - 3^3 = 229$, which is a prime number!

The Strange Challenge with the Number 44

There are times when numbers present a challenge that is not typically mathematical but more or less logical. Here is one such challenge. The first 44 numbers can be partitioned into four sets so that any two members of each set form a unique sum that is not within that set. These sets are:

$$\{1, 3, 5, 15, 17, 19, 26, 28, 40, 42, 44\},$$
$$\{2, 7, 8, 18, 21, 24, 27, 33, 37, 38, 43\},$$
$$\{4, 6, 13, 20, 22, 23, 25, 30, 32, 39, 41\}, \text{ and}$$
$$\{9, 10, 11, 12, 14, 16, 29, 31, 34, 35, 36\}.$$

The Triangular Number 45

The number 45 is a triangular number (see page 94) which produces the following sum: $0 + 1 + 2 + 3 + 4 + 5 + 6 + 7 + 8 + 9 = 45$. This triangular number can also be written as a sum of two squares as follows: $6^2 + 3^2 = 36 + 9 = 45$. Quite unexpectedly, we can subtract the first five powers of 2 from the number 45 to produce five prime numbers, as we show here:

$$45 - 2 = 43,$$
$$45 - 2^2 = 41,$$
$$45 - 2^3 = 37,$$
$$45 - 2^4 = 29,$$
$$45 - 2^5 = 13.$$

Using Prime Numbers to Exhibit the Composite Number 48

The composite number 48 is an even number that can be expressed as a sum of two prime numbers in five different ways, as we show here:

$$5 + 43 = 48,$$
$$7 + 41 = 48,$$
$$11 + 37 = 48,$$
$$17 + 31 = 48,$$
$$\text{and } 19 + 29 = 48.$$

The number 48 is a sum of four distinct primes, namely, 5, 7, 17 and 19, where the sum of any three of these four prime numbers yields a prime number, such as, for example, $5 + 7 + 17 = 29$, or $5 + 17 + 19 = 43$, and so on.

A Strange Relationship of the Square Root of 49

Consider $49^2 = 2401$, where, surprisingly, the sum of its digits $2 + 4 + 0 + 1 = 7$, which also is the square root of 49. The number 49 is comprised of digits which are each perfect squares. Curiously, $49^3 = 117{,}649$, where the last two digits are 49.

Representing as the Sum of Squares the Number 50

The number 50 can be expressed as a sum of two squares in two different ways, namely, $1^2 + 7^2 = 50$ and $5^2 + 5^2 = 50$. The number 50 can also be expressed as the sum of three squares, as $3^2 + 4^2 + 5^2 = 50$. Even more impressive is that the number 50 can also be expressed as the sum of four squared numbers, as $1^2 + 2^2 + 3^2 + 6^2 = 50$.

And to include the role of primes with the number 50, we see that we can use all the initial primes exactly once to express the number 50 as the sum of these primes in two different ways: $2 + 5 + 7 + 17 + 19 = 50 = 3 + 11 + 13 + 23$.

The Constant Reappearance of the Number 53

Weird things can be found in mathematics. For example, when we raise the number 53 to the seventh power, we find that $53^7 = 1{,}174{,}711{,}139{,}837$. The amazing result is that the sum of the digits of this number is $1 + 1 + 7 + 4 + 7 + 1 + 1 + 1 + 3 + 9 + 8 + 3 + 7 = 53$.

When we raise 53 to the third power, $53^3 = 148877$, and the sum of the digits $1 + 4 + 8 + 8 + 7 + 7 = 35$, which is the reverse of 53.

The number 53 can be seen as the average of three consecutive double-digit primes:

$$\frac{47+53+59}{3}=53,$$

and also, the average of seven consecutive primes:

$$\frac{41+43+47+53+59+61+67}{7}=53.$$

When we search for patterns amongst numbers, we can come up with some that are rather strange and somewhat entertaining. This is the case with the number 53, a two-digit prime number where the sum of the digits and the difference of the digits have a unique relationship. The sum of its digits $5 + 3 = 8 = 2^3$, and the number 2 is the difference of its digits, as $5 - 3 = 2$.

The Tricky Number 55

Here is a cute "trick" with the number 55. Consider the sum of the cubes of the digits $5^3 + 5^3 = 250$, and then the sum of the digits of 250 raised to the third power is $2^3 + 5^3 + 0^3 = 133$, and then when we get the sum of the cubes of the digits of 133, we essentially get back to the starting point, 55 as $1^3 + 3^3 + 3^3 = 55$. Notice how the sum of the cubes of the interim results takes us back to the number 55, which was the initial number.

The number 55 allows for a very unusual of products as we can see in Figure 1-13:

55×91 = 5005	55×101 = 5555
55×93 = 5115	55×103 = 5665
55×95 = 5225	55×105 = 5775
55×97 = 5335	55×107 = 5885
55×99 = 5445	55×109 = 5995

Figure 1-13

Surprises of the Number 59

Some unexpected patterns can leave the observer in awe. One such involves the number 59, as it is the smallest number that when divided by 2, 3, 4, 5, and 6 results with remainders of 1, 2, 3, 4, and 5, respectively. The number 59 is also the smallest power of 2 that results in a number where every digit appears multiple times, as we can see with $2^{59} = 576{,}460{,}752{,}303{,}423{,}488$. This is just a cute observation with no particular significance from a mathematical standpoint.

Here is a cute entertainment with the number 59. We have $59^2 = 3481$, then using the digits of the number 3481, we can create a relationship with the same order of the digits: $3^4 = 81$. Discoveries like this give one the opportunity to play with numbers.

Some Peculiarities of the Number 61

We can always find unusual relationships with numbers. One such is that the number 61 is the sum of the first three prime numbers that have a units digit of 7, namely, $7 + 17 + 37 = 61$. The decimal expansion of the reciprocal of the number 61 is peculiar. Although this hardly has any practical use, it is still worthy of note. The repeating part of the decimal of $\frac{1}{61}$ consists of 60 digits, where each digit 0 through 9 appears at least 6 times. Here is the expansion, which then repeats:

$$\frac{1}{61} = 0.\overline{016393442622950819672131147540983606557377049180327868852459}....$$

The number 61 is also the sum of 2 consecutive squares: $5^2 + 6^2 = 25 + 36 = 61$. However, the number 61 is also the smallest prime number that can be expressed as a square in 1, 2, 3 or 4 ways: $61^2 = 60^2 + 11^2$

$$61^2 = 52^2 + 24^2 + 21^2,$$
$$61^2 = 56^2 + 22^2 + 10^2 + 1^2 = 3{,}721.$$

We accentuate the prime numbers to use them in numerous ways for entertainment. The number 61 can be described as the smallest prime number that can be written as the sum of a prime number of prime numbers, each taken to a prime power in a prime number (three) of ways:

$$61 = 2^2 + 2^3 + 7^2, \quad 61 = 2^2 + 5^2 + 2^5, \quad \text{and } 61 = 3^2 + 5^2 + 3^3.$$

As we will see later in this chapter, the number 1089 has very special properties. However, before we get to that point, we would like to show that the number 61 is also related to the number 1089 in a most unusual fashion:

$$\frac{1098}{1+0+9+8}=61.$$

The Kaprekar Number 63

We will encounter the various discoveries of the Indian mathematician Dattatreya Ramchandra Kaprekar (1905–1986) throughout the book. (See page 94.) One such discovery that bears his name is the Kaprekar number: 63. This number lends itself to the Kaprekar process—reverse the digits and subtract—which we show as follows: $63 - 36 = 27$, then $72 - 27 = 45$, then $54 - 45 = 09$, then $90 - 09 = 81$, and then $81 - 18 = 63$, which is the number we started with. Furthermore, the number 63 can be represented as the sum of consecutive powers of 2, as $2^0 + 2^1 + 2^2 + 2^3 + 2^4 + 2^5 = 63$.

The Uniqueness of the Number 65

The number 65 is the only number that can generate a square number by adding its reversal: $65 + 56 = 121 = 11^2$, and can generate a square number by subtracting its reversal: $65 - 56 = 9 = 3^2$.

There is also a Pythagorean triple where the hypotenuse is 65 and one of its legs is its reversal 56. That triple is (33, 56, 65) since $33^2 + 56^2 = 65^2$. An analogous situation can be seen with the Pythagorean triple (3333, 5656, 6565). Notice the symmetry and relationship that this triple has with the previous one. This can of course be justified in that $3333^2 + 5656^2 = 43{,}099{,}225 = 6565^2$.

Some Unique Features of the Number 66

If one takes the sum of all the divisors of the number 66, a perfect square will be obtained, as $1 + 2 + 3 + 6 + 11 + 22 + 33 + 66 = 144 = 12^2$.

The number 66 has a peculiar way of being represented as the sum of two prime numbers, each with 3 as the unit digit, in two different ways, as shown here: $66 = 13 + 53 = 23 + 43$.

The number 63 is also the smallest number that can be expressed as the sum of two prime numbers in six different ways: $66 = 5 + 61 = 7 + 59 = 13 + 53 = 19 + 47 = 23 + 43 = 29 + 37$.

A Peculiar Aspect of the Number 67

It is not always easy to come up with unusual relationships between numbers. One curious way to represent the number 67 is $2^6 + 2^1 + 2^0 = 26 + 21 + 20 = 67$. Notice how the first representation (including the exponents) relates to the second representation. Again, this shows a certain creativity in mathematics.

The Digits Represented by the Number 69

The number 69 is the only number whose square and cube use all the numbers from 0 to 9 exactly once, as we can see: $69^2 = 4{,}761$, and $69^3 = 328{,}509$.

Curiosities with the Number 71

First of all, the number 71 is a factor of the sum of all the prime numbers less than 71, that is: $2 + 3 + 5 + 7 + 11 + 13 + 17 + 19 + 23 + 29 + 31 + 37 + 41 + 43 + 47 + 53 + 59 + 61 + 67 = 71 \times 8$. A true curiosity emerges when we consider $71^3 = 357{,}911$, and we inspect the digits to find that all the primes between 3 and 11 are represented in order. We can also arrive at the number 71 by taking the sum of three consecutive prime numbers, namely, $71 = 19 + 23 + 29$. Not to ignore the composite numbers, we find that the number 71 can be expressed in two ways as the sum of consecutive composite numbers as follows: $71 = 22 + 24 + 25$, and $71 = 35 + 36$.

There are three prime numbers in the 70s, and they are 71, 73, 79. Their sum is 223, which is also a prime number. The reversals of these three prime numbers, namely, 17, 37, 97, have a sum of 151, which is also a prime number.

Primes Generated by the Composite Number 72

The number 72 can be expressed as the sum of four consecutive prime numbers: $72 = 13 + 17 + 19 + 23$. It is also the sum of six consecutive prime numbers: $72 = 5 + 7 + 11 + 13 + 17 + 19$.

The number 72 is the smallest number that can be expressed as a sum of two prime numbers, each of which has a units digit of 1, as shown: $72 = 11 + 61$, and $72 = 31 + 41$.

The number 72 is also the smallest number that can be expressed as a difference of two squares of consecutive prime numbers in two different ways: $72 = 19^2 - 17^2$, and $72 = 11^2 - 7^2$.

With the following two sets of numbers, the sum of the numbers in each set is 72 and the sum of any three of the numbers in each of the sets will always be a prime number. The two sets are $\{5, 11, 13, 43\}$ and $\{11, 13, 19, 29\}$. This next property is not very easily discoverable, but it is worthy of notice. The number 72 is the smallest number whose fifth power is equal to the sum of five other fifth powers: $72^5 = 19^5 + 43^5 + 46^5 + 47^5 + 67^5$.

Mixed with Prime Numbers Is the Number 73

The prime number 73 is composed of prime digits and its reversal is also a prime number, 37. While we are on the number 73 and its reversal 37, we can generate another prime number with these numbers: $(73 \times 37) + 7^3 + 3^7 + (7 \times 3) + (3 \times 7) = 2{,}593$, which is the 378th prime number.

Here is a cute use of the number 73. The 73rd triangular number is equal to $73 \times 37 = 2{,}701$. By the way, the formula for finding the nth triangular number is $\frac{n(n+1)}{2}$. (See page 94.)

Representing the Number 77

The number 77 is the sum of the first eight prime numbers: $77 = 2 + 3 + 5 + 7 + 11 + 13 + 17 + 19$.

The number 77 can also be expressed as a sum of three consecutive squares: $77 = 4^2 + 5^2 + 6^2$.

If one takes all the palindromic numbers from 1 through 77 and places them together to form the number 12,345,678,911,223, 344,556,677, the result is a prime number.

The number 77 will appear as a sum of the digits of a number formed by lots of 7s in the following way: $7^7 + 77^7 + 777^7 + 2 =$ 170,980,732,128,390,323,351, whose digit sum is 77.

The Curiosities of the Number 79

The 79th triangular number is 3,160, which is comprised of four digits, each of which is a triangular number. The number 79 is the 22nd prime number. Its reversal, 97, is also prime. Both of these numbers can be expressed as the sum of three prime numbers that are also reversable primes: $79 = 11 + 31 + 37$, and $97 = 11 + 13 + 73$. The number 79 can be arrived at by adding the product of its digits to the sum of its digits as follows: $(7 \times 9) + (7 + 9) = 63 + 16 = 79$.

Another clever way of representing the number 79 using only prime numbers is as follows:

$$2^7 - 7^2 = 79.$$

The number 79 is the average of the four prime numbers on either side of 79, namely, 73, 71, 83, and 89. Symbolically,

$$\frac{73+71+83+89}{4} = 79.$$

The number 79 is also a happy number, since $7^2 + 9^2 = 49 + 81 = 130$, and $1^2 + 3^2 = 10$, and $1^2 + 0^2 = 1$.

Interestingly, the number 79 lends itself to a very nice little trick. Select any two two-digit numbers that differ by 2 and with different tens digits. Then reverse each of the two numbers and subtract the smaller from the larger. Surprisingly, the result will always be 79. For example, consider the selected two numbers 39 and 41, which differ by 2. When we subtract their reversals: $93 - 14 = 79$.

Unusual Appearances of the Number 81

The reciprocal of 81 is $\frac{1}{81} =$ 0.**012345679** 012345679 012345679 012345679 . . . , where the digits are in order without the number 8.

We can also lead up to the number 81 with a very curious pattern as follows:

$$1 = 3^0 = 1$$
$$2 + 3 + 4 = 3^2 = 9$$
$$5 + 6 + 7 + 8 + 9 + 10 + 11 + 12 + 13 = 3^4 = \mathbf{81}$$
$$14 + 15 + 16 + 17 + 81 + \cdots + 36 + 37 + 38 + 39 + 40 = 3^6 = 729.$$

Another Happy Number Is 82

The number 82 is considered a "happy number," which, as mentioned earlier, is a number that when you continuously take the sum of the squares of the digits eventually will reach the number 1. For the number 82, the process goes as follows: $8^2 + 2^2 = 68 \to 6^2 + 8^2 = 100 \to 1^2 + 0^2 + 0^2 = 1$.

Using Prime Numbers to Express the Prime Number 83

The number 83 is the 23rd prime number and can be expressed as a sum of three consecutive prime numbers, namely, $23 + 29 + 31 = 83$. The number 83 can also be expressed as the sum of five consecutive prime numbers, namely, $11 + 13 + 17 + 19 + 23 = 83$.

The smallest prime number where the sum of its digits equals 83 is the prime number 3,999,998,999.

The number 83 is equal to the sum of the squares of the first three consecutive odd prime numbers: $83 = 3^2 + 5^2 + 7^2$.

The number 83 can also be obtained by taking the sum of the first three prime numbers whose units digit is 1, as we see with the following: $83 = 11 + 31 + 41$.

The number 83 is the only two-digit prime number which is the average of three reversible prime numbers:

$$\frac{73+79+97}{3} = 83.$$

The average of all the prime numbers less than or equal to 83 is the reversal of 83, namely 38.

$$\frac{2+3+5+7+11+13+17+19+23+29+31+37+41+43+47+53+59+61+67+71+73+79+83}{23}$$
$$= \frac{874}{23} = 38.$$

The square of the number 83 results in a number whose value doesn't change when it's flipped over. That is, $83^2 = 6889$.

The Number 84

Although the number 84 is not a prime number, it is the sum of two twin prime numbers: $41 + 43 = 84$. The number 84 is the smallest number that can be expressed as a sum of two prime numbers in eight different ways, and they are as follows: $(5 + 79)$, $(11 + 73)$, $(13 + 71)$, $(17 + 67)$, $(23 + 61)$, $(31 + 53)$, $(37 + 47)$, $(41 + 43)$.

The number 84 is also the smallest number that can be expressed as the sum of four different primes, where the sum of any three primes within each set is also a prime number. These four sets of prime numbers are: $\{5 + 13 + 23 + 43\}$, $\{11 + 13 + 17 + 43\}$, $\{11 + 13 + 23 + 37\}$, $\{13 + 17 + 23 + 31\}$. For example, if we take the first set listed, we can see that the sum of any three of the numbers in the group yields a prime number: $5 + 13 + 23 = 41$, $13 + 23 + 43 = 79$, $5 + 23 + 43 = 71$, $5 + 13 + 43 = 61$.

Uniqueness of 2 to the Power of the Number 86

It is believed that $2^{86} = 77{,}371{,}252{,}455{,}336{,}267{,}181{,}195{,}264$ is the largest power of 2 that does not contain a digit 0. When we go to the next power of 2, that is, 2^{87} there are three zeros included in the number: $2^{87} = 154{,}742{,}504{,}910{,}672{,}534{,}362{,}390{,}528$.

By the way, as we noted earlier with the number 82, the number 86 is also a happy number: $8^2 + 6^2 = 64 + 36 = 100 \rightarrow 1^2 + 0^2 + 0^2 = 1$.

Generating Prime Numbers with the Composite Number 87

The number 87 is a composite number as it has two factors, 3 and 29. However, if we place each of the single-digit prime numbers 2, 5, and 7 between the two numbers 8 and 7, a prime number is created: 8**2**7 is the 144th prime number, 8**5**7 is the 148th prime number, and 8**7**7 is 151st prime number.

Another clever pattern to generate prime numbers can emanate from the number 87 in the following way. The two factors of 87, 3 and 29, are prime numbers that can generate prime numbers as follows: $87^2 - 3^2 - 29^2 = 6719$, which is the 867th prime number, and $87^2 + 3^2 + 29^2 = 8419$, which is the 1052nd prime number.

The Toughest Path to a Palindrome Is the Number 89

There is a nice technique for generating palindromic numbers using most starting numbers. One can arrive at a palindromic number by taking the given number and adding it to his reversal and continuing this process until a palindromic number is reached. There are times when this can be accomplished in very few steps. For example, if one were to begin with a number 86, the palindrome would be reached in three steps: $86 + 68 = 154$, $154 + 451 = 605$, and $605 + 506 = 1111$, a palindromic number. Were one to do this with the number 89 it would require 24 steps to reach the palindrome, which is probably the maximum number of steps. An ambitious reader might try it to verify the 24 steps.

Cute patterns for 89 are that $89 = 8 \times 9 + (8 + 9)$, and that $89 = 9 + 8 + 7 + 6 + 5 + 4 + 3 + 2 + 1 + 2 + 3 + 4 + 5 + 6 + 7 + 8 + 9$. Notice the palindromic pattern.

If we were to test the number 89 to see if it is a happy number, (see page 100) we would use the process described for the number 82 (page 50), and rather than reach the number 1, the number 89 would reappear and create a loop. The number 89 joins the number 16 as the only two looping numbers that occur if you take the sum of the squares of the digits and continue this process. For example, suppose the number 16 is selected as the starting number: $1^2 + 6^2 = 37$, then $3^2 + 7^2 = 58$, then $5^2 + 8^2 = 89$, then $8^2 + 9^2 = 145$, $1^2 + 4^2 + 5^2 = 42$, then $4^2 + 2^2 = 20$, then $2^2 + 0^2 = 4$, then $4^2 = 16$, and if you continue the process you will get back to the number, 16. Now we consider the number 16 again and form another loop, as we can see with $1^2 + 6^2 = 37$, then $3^2 + 7^2 = 58$, then $5^2 + 8^2 = 89$. These numbers which end up with 89 and 16 are called "unhappy numbers," as opposed to the happy numbers which end up with 1.

Just for fun, it might be interesting to know that, although it may be difficult to count them, there are exactly 1000 prime numbers between 1 and $89^2 = 7921$.

A Cute Arrangement of Numbers Generating the Number 90

There are times when mysterious relations with numbers can be enriching. For example, there is an unusual relationship between two groups of prime numbers in specific order, where the first member of each group, the second member of each group, the third member of each group, and so on, will have a sum of 90. The two groups are $(17, 19, 23, 29, 31, 37)$ and $(73, 71, 67, 61, 59, 53)$. For example, taking the sum of the last member of each of the two groups: $37 + 53 = 90$. For another example, the fourth member of each group yields $29 + 61 = 90$.

Clever Ways of Generating the Number 91

The number 91 can be reached in a variety of clever ways. As we exhibit these varieties you may notice the relationship of the members of the sums.

$$91 = 1 + 2 + 3 + 4 + 5 + 6 + 7 + 8 + 9 + 10 + 11 + 12 + 13.$$
$$91 = 1 + 6 + 12 + 18 + 24 + 30$$
$$91 = 1^2 + 2^2 + 3^2 + 4^2 + 5^2 + 6^2$$
$$91 = 1^2 + 4^2 + 5^2 + 7^2$$
$$91 = 1^2 + 3^2 + 9^2.$$

The number 91 can also be expressed as a sum of two cubes and the difference of two cubes, as we see with $91 = 4^3 + 3^3$, and $91 = 6^3 - 5^3$.

Curiosities with the Number 97

The number 97 and its reversal 79 are both prime numbers. The next number after 97 with the units digit of 7 is the number 107, which is also a prime number, as is its reversal, 701. Coincidences in mathematics are surely noteworthy. For example, as we consider the number 97, we find that there are 97 prime numbers between 107 and 701.

Curiously, we can insert several 0s between the two digits of 97 to generate some prime numbers as follows: 907, 9007, 90007, and 900007, which are four prime numbers.

Cute Properties of the Number 99

The number 99 is a Kaprekar number, since $99^2 = 9801$ and $98 + 01 = 99$.

Whenever we explore a number 1 less than a power of 10, we can expect some strangeness in the base 10. Considering the reciprocal of 99, there we also find a very curious pattern:

$$\frac{1}{99} = .01010101010101010101010101\ldots.$$

Reaching the Number 100

As would be expected, the number 100 can be expressed in various interesting ways:

- As the square of the sum of the first four natural numbers: $100 = (1 + 2 + 3 + 4)^2$.
- As the sum of the cubes of the first four natural numbers: $100 = 1^3 + 2^3 + 3^3 + 4^3$.
- As the sum of the first nine prime numbers: $100 = 2 + 3 + 5 + 7 + 11 + 13 + 17 + 19 + 23$.
- As the sum of 2 to the sixth power and 6 to the second power: $2^6 + 6^2 = 64 + 36$.
- By using the digits from 1 to 9 to get $100 = 1 + 2 + 3 + 4 + 5 + 6 + 7 + (8 \times 9)$, or using only three operational signs: $100 = 123 - 45 - 67 + 89$.

A motivated reader may develop other ways of reaching the number 100 through a variety of arithmetic operations.

The Ubiquitous Number 101

The number 101 is the 26th prime number, and it is also a palindromic number.

Once more, we see a curiosity with a number close to the base 10 power and that is the reciprocal of 101:

$$\frac{1}{101} \approx 0.00990099009900990099009900990099009900990099\ldots$$

which is quite unique among prime numbers.

The number 101 is the largest known prime 1 greater than a power of 10.

The number 101 can be expressed as a sum of five consecutive prime numbers:

$$101 = 13 + 17 + 19 + 23 + 29.$$

The number 101 can be obtained by taking the sum of the products of all combinations of the first 4 prime numbers: $101 = (2 \times 3) + (5 \times 2) + (2 \times 7) + (3 \times 5) + (3 \times 7) + (5 \times 7)$.

The number 101 is the first of 21 prime numbers where, when the center digit is removed, the remaining number is still a prime number. For example, here are a few of these prime numbers that convert to a prime number when the center digit is removed: $(101 \rightarrow 11)$, $(103 \rightarrow 13)$, $(107 \rightarrow 17)$, ... $(193 \rightarrow 13)$, $(197 \rightarrow 17)$ and $(199 \rightarrow 19)$.

Reversals with the Number 103

There are times when it is a bit of a mystery how a prime number can generate another prime number. One such is to generate a prime number from the prime number 103 and involves a very unusual procedure. We will take the reverse of the product of the number 103 and its reverse 301. That is, $103 \times 301 = 31003$, and its reverse is 30013, which is a prime number.

Coincidental Digit Sums for the Number 107

Consider the sum of the digits of each of the following:

$$107^{11} = 21{,}048{,}519{,}522{,}998{,}348{,}950{,}643;$$
$$107^{13} = 240{,}984{,}500{,}018{,}808{,}097{,}135{,}911{,}707; \text{ and}$$
$$107^{15} = 2{,}759{,}031{,}540{,}715{,}333{,}904{,}109{,}053{,}133{,}443.$$

Surprisingly, each of these three huge numbers has a digit sum equal to 107.

Unexpected Reversals for the Number 113

The number 113 is the smallest three-digit prime number whose digits can be arranged in any order and still result in a prime number. That is, 131, and 311 are prime numbers.

A rare curiosity occurs when we take $113^2 = 12{,}769$ and then take the square of the reversal $311^2 = 96{,}721$, which amazingly is a reversal of the previous square.

Consider a number formed by the consecutive natural numbers up to and including the number 13, which will be: 12,345,678,910,111,213 (or better seen as 1 2 3 4 5 6 7 8 9 10 11 12 13). Strangely enough, this huge number has 113 as a factor; that is,

$$\frac{12{,}345{,}678{,}910{,}111{,}213}{113} = 109{,}253{,}795{,}664{,}701.$$

The history of mathematics is full of estimates of the value of π. We know that the estimates can go on infinitely long, such as $\pi \approx 3.14159265358979323846264338327950288 4197\ldots$. We can also get a good estimate of the value of π by taking the quotient $\frac{355}{113} \approx$ **3.141592** 920353982300884955752212389380530973 45…, where we notice we have the value accurate to six decimal places. This is quite astonishing and surprising.

At times it could be interesting to challenge our arithmetic skills. For example, it is possible to create the number 113 from seven different sums using only the first 22 prime numbers in seven different ways:

$$113 =$$
$$2 + 19 + 31 + 61 =$$
$$3 + 37 + 73 =$$
$$5 + 41 + 67 =$$
$$7 + 47 + 59 =$$
$$11 + 23 + 79 =$$
$$13 + 29 + 71 =$$
$$17 + 43 + 53.$$

The sum of the digits of the number 113 and the product of digits are both prime numbers, which are $1 + 1 + 3 = 5$, and $1 \times 1 \times 3 = 3$, respectively. These results are twin primes, 3 and 5.

The number 113 is the smallest prime number that can be expressed as a sum of three fourth powers of prime numbers: $2^4 + 2^4 + 3^4 = 113$.

Strange Coincidences with the Number 118

In mathematics we have opportunities to create uncommon relationships. The number 118 can be written as the sum of three numbers in four different ways, where the product of those three numbers, in each case, is the same, namely, 37,800. The four ways to write the sums are: **118** $= 15 + 40 + 63$, **118** $= 14 + 50 + 54$, **118** $= 18 + 30 + 70$, and **118** $= 21 + 25 + 72$. Now here comes the amazing part. The product of each group of three numbers is as follows: $15 \times 40 \times 63 = 14 \times 50 \times 54 = 18 \times 30 \times 70 = 21 \times 25 \times 72 = 37{,}800$.

Another unusual aspect involving the number 118 is that it one of two consecutive composite numbers, 118 and 119, whose reversals, 811 and 911, are both prime numbers.

Prime Numbers Generate the Number 119

The number 119 can be expressed as a sum of five consecutive primes as follows: $119 = 17 + 19 + 23 + 29 + 31$. Additionally, 119 can also be expressed as a sum of seven primes as follows: $119 = 7 + 11 + 13 + 17 + 19 + 23 + 29$.

The number 119 can be expressed as a product of two primes whose units digit is a 7, namely, $119 = 7 \times 17$.

The Divisors of the Number 120

The number 120 has more divisors (16) than any smaller positive number. These divisors are as follows: 1, 2, 3, 4, 5, 6, 8, 10, 12, 15, 20, 24, 30, 40, 60, 120. Surprisingly, the sum of these divisors is $3 \times 120 = 360$.

Some Unique Features of the Number 121

The number 121 is the only palindromic number that can be expressed as the sum of increased powers of a prime number, such as $121 = 3^0 + 3^1 + 3^2 + 3^3 + 3^4$.

The number 121 can be expressed in three different ways as the sum of a prime number and its reversal: $121 = \mathbf{29} + 92$, $121 = \mathbf{83} + 38$, and $121 = \mathbf{47} + 74$.

The famous French mathematician Pierre de Fermat (1607–1665) conjectured that 121 is one of two square numbers that when increased by 4 will become a perfect cube ($125 = 5^3$). The other such number is 4, since $8 = 2^3$.

Reversals of Squares and Cubes Generate the Number 127

We can arrive at the number 127 in a rather strange way. If we take the sum of the reversals of the first five numbers squared (1 ,4, 9, 16, 25), we get $1 + 4 + 9 + 61 + 52 = 127$. We can also attain 127 by taking the sum of the reversals of the first four numbers cubed (1, 8, 27, 64) to get $1 + 8 + 72 + 46 = 127$.

We should appreciate that the number 127 is equal to the sum $2^0 + 2^1 + 2^2 + 2^3 + 2^4 + 2^5 + 2^6$.

The number 127 is the smallest prime number that can be written as a sum of the first two or more odd prime numbers, as we can see, where 127 = 3 + 5 + 7 + 11 + 13 + 17 + 19 + 23 + 29. A motivated reader should come up with other schemes to represent the number 127.

When we talk about coincidences, it's difficult to top this one. The 127th prime number is 709, and the 127th happy number is 907. Not only are they reversals of one another, but they are also both prime numbers.

The number 127 has gained fame for its position of disproving a mathematician's conjecture. The French mathematician Alphonse de Polignac (1817–1890) stated that "every odd number greater than 1 can be expressed as the sum of a power of 2 and a prime number." If we inspect the first few cases, we find that this appears to be a true statement. However, as you will see in Figure 1-14, while this proposition holds true for the odd numbers from 3 through 125, it is *not* true for 127; after which it continues to hold true again for a while.

Odd number	Sum of a power of 2 and a prime number
3	$= 2^0 + 2$
5	$= 2^1 + 3$
7	$= 2^2 + 3$
9	$= 2^2 + 5$
11	$= 2^3 + 3$
13	$= 2^3 + 5$
15	$= 2^3 + 7$
17	$= 2^2 + 13$
19	$= 2^4 + 3$
…	…
51	$= 2^5 + 19$
…	…
125	$= 2^6 + 61$
127	**= ?**
129	$= 2^5 + 97$
131	$= 2^7 + 3$

Figure 1-14

The next numbers that fail de Polignac's conjecture are 149, 251, 331, 337, 373, and 509, while another counterexample is 877.

Another Happy Number 129

The number 129 is also a happy number since $1^2 + 2^2 + 9^2 = 86$, and $8^2 + 6^2 = 100$, and $1^2 + 0^2 + 0^2 = 1$, which verifies the happy number status.

The number 129 is equal to the sum of the first 10 prime numbers:

$$2 + 3 + 5 + 7 + 11 + 13 + 17 + 19 + 23 + 29 = 129.$$

The number 129 is also the smallest number that can be expressed as the sum of three squares in four different ways: $129 = 1^2 + 8^2 + 8^2$, $129 = 2^2 + 2^2 + 11^2$, $129 = 4^2 + 7^2 + 8^2$, and $129 = 10^2 + 5^2 + 2^2$.

Strange Curiosities Involving Number 131

If we place a 7 on either side of the palindromic number 131, we get the number 71,317, which is also a palindrome and also a prime number. The same thing is true if we tag a 1 to each side of the number 131 to get 11,311, which again is both a palindrome and a prime number.

Perhaps also impressive is that the number 131 can be expressed as follows:

$$131 = (1^0 + 3^0 + 1^0) + (1^1 + 3^1 + 1^1) + (1^2 + 3^2 + 1^2) + (1^3 + 3^3 + 1^3) + (1^4 + 3^4 + 1^4).$$

A further curiosity regarding the number 131 is that it is the sum of the odd numbers that have no even digits from 1 to 31, as shown here: $131 = 1 + 3 + 5 + 7 + 9 + 11 + 13 + 15 + 17 + 19 + 31$.

In mathematics we sometimes look for coincidences. Consider the first few digits of the value of π, which are 3.14159. If we separate these digits into pairs and add those numbers: $31 + 41 + 59$, we get a sum of 131.

Unique Unusualness of the Number 132

The number 132 and its reversal 231 are both divisible by 11. Not only is this highly unusual, but the two quotients produced when dividing by 11, that is, $\frac{132}{11}=12$ and $\frac{231}{11}=21$, are also reversals of one another. This is even more unusual.

The number 132 is the smallest number that can be expressed as the sum of the two-digit numbers formed by all variations of its three digits. That is, $12 + 13 + 21 + 23 + 31 + 32 = 132$.

Strange Representations of Number 135

The number 135 can be expressed in a rather strange fashion: $135 = (1 + 3 + 5) \times (1 \times 3 \times 5)$, or as $135 = 1^1 + 3^2 + 5^3$. Furthermore, the number 135 can be expressed as the sum of consecutive squares: $135 = 3^2 + 4^2 + 5^2 + 6^2 + 7^2$.

An Unexpected Sum of the Number 136

The number 136 is equal to the sum of the first 16 positive integers, namely,

$$1+2+3+4+5+6+7+8+9+10+11+12+13+14+15+16=136.$$

Palindromes about the Number 137

The reciprocal of 137, namely, $\frac{1}{137} = 0.\overline{00729927}00$, produces a repeating *palindromic* decimal, which is most unusual. While considering palindromes, the number 137 is a factor of the palindrome 123,456,787,654,321, which is equal to $137 \times 901{,}144{,}435{,}433$. The number 137 is also a factor of the repunits number 11,111,111 since $137 \times 81{,}103 = 11{,}111{,}111$.

The number 137 is the only three-digit prime number which is a factor of a number whose digits follow the pattern xxxxxYYxxYYxxxxx. For example, 4,444,455,445,544,444 = 137 × 32,441,280,624,412. A motivated reader might want to try this with other such numbers.

Prime Numbers Generated by Number 139

The number 139 can be used as exponents to generate other prime numbers like so: $2^1 + 2^3 + 2^9 - 1 = 521$, which is a prime number. Also, $2^1 + 2^3 + 2^9 + 1 = 523$, another prime number. We can extend this as follows: $2^1 + 2^3 + 2^9 + 139^2 = 19{,}843$, which is a prime number as well.

The product of the digits of the prime number 139 is a cube of a prime number, namely, $1 \times 3 \times 9 = 27 = 3^3$.

Once again, we have a happy number with the number 139, since $1^2 + 3^2 + 9^2 = 91$, $9^2 + 1^2 = 82$, $8^2 + 2^2 = 68$, $6^2 + 8^2 = 100$, and $1^2 + 0^2 + 0^2 = 1$. Thus, 139 qualifyies as a happy number.

Squares Generate the Number 140

The number 140 can be expressed as a sum of the squares of the first seven natural numbers:

$$1^2 + 2^2 + 3^2 + 4^2 + 5^2 + 6^2 + 7^2 = 140.$$

Curiosities of the Number 143

The number 143 can be used in a rather strange way to generate a prime number as follows:

$$\frac{143143}{143^2} = \frac{143143}{20449} = 7.$$

The difference between $3^4 + 4^4 + 5^4 + 6^4$ and 7^4 is 143, since $2258 - 2401 = 143$.

Unusual Relationship with the Reversal of the Number 144

We know that $144 = 12^2$. Curiously, if you reverse 144, the square will also be reversed, since $441 = 21^2$.

Primes Surprisingly Generated by the Prime Number 149

The number 149 lends itself to an unusual prime number construction, where the number is formed by the number of digits, which is determined by the value of the digit. The basis is the number 149, and the number produced in this unusual fashion has one digit 1, four digits of 4, and nine digits of 9, resulting in the prime number 14,444,999,999,999.

We can use a clever little trick to create new prime numbers from the number 149 by placing a 0 between pairs of digits to get the numbers 1049 and 1409, which are both prime numbers.

The Largest Gap Between Twin Primes Is the Number 150

Twin prime numbers, which are consecutive prime numbers, occur rather often at the start of our list of prime numbers but grow more infrequent the further we go. The largest gap between pairs of twin primes between 1 and 1000 is 150 numbers, which occurs between the pairs of twin primes (659, 661) and (809, 811).

The Tricky Number 153

Not only is the number 153 the 17th triangular number, but it also is equal to the sum of the first 17 integers: $1 + 2 + 3 + 4 + 5 + 6 + 7 + 8 + 9 + 10 + 11 + 12 + 13 + 14 + 15 + 16 + 17 = 153$, which is quite an unusual coincidence.

Furthermore, there is a clever trick that can be done with the number 153. Select any integer that is divisible by 3 and then take the sum of the cubes of its digits. Continue to do that with each result and you will end up with the number 153. For example, suppose we begin with the number 297. The sum of the cubes of its digits is $2^3 + 9^3 + 7^3 = 1080$. Continuing this process: $1^3 + 0^3 + 8^3 + 0^3 = 513$, then $5^3 + 1^3 + 3^3 = 153$.

It is just a coincidence that the number 297 ended up as 153. However, taking the sum of the cubes of the digits of any number and repeating will get you into a cycle. For example, the number 437 takes nine steps to reach the number 371, and taking the sum of the cubes of the digits of 371 continue to produce 371. As another example, the number 856 reaches the number 370 in six steps and then continues to repeat that number. This is just a nice little trick to show the amazements hidden in mathematics.

Sum of Consecutive Primes Emits the Number 160

The number 160 can be expressed as the sum of the first 11 prime numbers as follows:

$$160 = 2 + 3 + 5 + 7 + 11 + 13 + 17 + 19 + 23 + 29 + 31.$$

Perfect Numbers Generate the Number 168

The first two perfect numbers in our number system are 6 and 28, since each of these numbers is equal to the sum of their factors. However, the number 168 is equal to the product of these two perfect numbers: $6 \times 28 = 168$.

There are 168 prime numbers less than 1000, that is, between 2 and 997.

A Rare Curiosity with the Number 169

Amazingly, the number 169 is a square number ($169 = 13^2$) that, when reversed, is also a square number ($961 = 31^2$).

A curiosity involving the number 169 occurs when we replace the number with one that repeats the digit 1 once, the digit 6 six times, and the digit 9 nine times. The result is a prime number: 1,666,666,999,999,999. The same is true with the reverse number 961, where 9,999,999,996,666,661 is a prime number. When the flipped number 691 is treated analogously it also becomes a prime number: 6,666,669,999,999,991.

Prime Numbers Generating the Number 177

The number 177 is the sum of all the prime numbers from 2 to 47, that is, $2 + 3 + 5 + 7 + 11 + 13 + 17 + 19 + 23 + 29 + 31 + 37 + 41 + 43 + 47 = 177$. The 15th $(1 + 7 + 7 = 15)$ prime number is 47, which was the last prime we used in this sum.

Prime Numbers Involving the Number 179

A prime number can be generated with the number 179 and its reversal 971 as follows:

$$(179 \times 971) + 179 + 971 = 174{,}959, \text{ which is a prime number.}$$

The sum of all double-digit prime numbers with consecutive digits in ascending order is 179, as shown here: $179 = 23 + 67 + 89$.

Digits Generating the Number 181

The number 181 is equal to the sum of the digits of the first 23 prime numbers, namely, from 2 to 83, shown here: $2 + 3 + 5 + 7 + 1 + 1 + 1 + 3 + 1 + 7 + 1 + 9 + 2 + 3 + 2 + 9 + 3 + 1 + 3 + 7 + 4 + 1 + 4 + 3 + 4 + 7 + 5 + 3 + 5 + 9 + 6 + 1 + 6 + 7 + 7 + 1 + 7 + 3 + 7 + 9 + 8 + 3$.

The Happy Number 188

The number 188 is a happy number, since

$$1^2 + 8^2 + 8^2 = 129,\ 1^2 + 2^2 + 9^2 = 86,\ 8^2 + 6^2 = 100,$$

and finally, $1^2 + 0^2 + 0^2 = 1$. Thus, a happy number is achieved.

Another Happy Number 190

The number 190 is a happy number as we can see from the following: $1^2+9^2+0^2=82$, $8^2+2^2=68$, and $6^2+8^2=100$, which then leads to 1, making 190 a happy number.

The Squares Inside the Number 191

If we inspect the square of 191, we get $191^2=36{,}481$. Curiously, the two digits bookending either side of this result, 36 and 81, are both perfect squares. The middle number, 4, is also perfect square.

We now have a very unusual relationship where the number 191 is the first of four consecutive prime numbers (191, 193, 197, and 199) where the sum of the digits of each of these four prime numbers (11, 13, 17, and 19, respectively) are also consecutive prime numbers.

Yet Another Happy Number 192

The number 192 is also a happy number, since $1^2+9^2+2^2=86$, and $8^2+6^2=100$, which then leads to 1, making 192 a happy number.

There are occurrences in mathematics that are inexplicable. Consider the addition 192 + 384 = 576. You may ask, what is so special about this addition? Look at the outside digits (bold): **192** + **384** = **576**. They are in numerical sequence from left to right (1, 2, 3, 4, 5, 6), and then you reverse to get the rest of the nine digits (7, 8, 9). You might have also noticed that the three numbers are:

$$192=1\times 192,$$
$$384=2\times 192,$$
$$576=3\times 192.$$

The Tricky Number 193

The number 193 lends itself to an interesting manipulation. If we square 193, we get 37,249. If we attach onto this square the reverse number 94,273, we get the palindrome 3,724,9**94,273**, which is the product of one palindromic prime number and one palindromic non-prime number: $3{,}721{,}273 \times 1{,}001$.

The number 193 is a prime number as well as a happy number, which we can show by the following: $1^2 + 9^2 + 3^2 = 91$, $9^2 + 1^2 = 82$, $8^2 + 2^2 = 68$, and $6^2 + 8^2 = 100$, which then leads to 1, making 193 a happy number.

The Infamous Number 196

The number 196 is famous in mathematics for being the smallest number that through the reversal-addition process does not produce a palindromic number. (At least, to date it has not produced a palindromic number.) For example, the number 195 does produce a palindrome with continuous addition of reversals: $195 + 591 = 786$, $786 + 687 = 1473$, $1473 + 3741 = 5214$, and $5214 + 4125 = 9339$, which is a palindrome.

The Digits of the Number 197

The prime number 197 is the sum of all the two-digit prime numbers:

$$197 = 11 + 13 + 17 + 19 + 23 + 29 + 31 + 37 + 41 + 43 + 45 + 53 + 59 + 61 + 67 + 71 + 73 + 79 + 83 + 89 + 97.$$

The number 197 can also be obtained by taking the sum of the digits of all the two-digit prime numbers as follows: $197 = \mathbf{1} + \mathbf{1} + 1 + 3 + \mathbf{1} + \mathbf{7} + 1 + 9 + \mathbf{2} + \mathbf{3} + 2 + 9 + \mathbf{3} + \mathbf{1} + 3 + 7 + \mathbf{4} + \mathbf{1} + 4 + 3 + \mathbf{4} + \mathbf{7} + 5 + 3 + \mathbf{5} + \mathbf{9} + 6 + 1 + \mathbf{6} + \mathbf{7} + 7 + 1 + \mathbf{7} + \mathbf{3} + 7 + 9 + \mathbf{8} + \mathbf{3} + 8 + 9 + \mathbf{9} + \mathbf{7}$.

Furthermore, the prime number 197 is also the sum of the first 12 prime numbers:

$$197 = 2 + 3 + 5 + 7 + 11 + 13 + 17 + 19 + 23 + 29 + 31 + 37.$$

The Permuted Number 199

When the digits of the prime number 199 are permuted, each result is also a prime number, as 991 and 919 are both prime numbers.

When the number 199 is attached to the next prime number, 211, to get 199,211 or, in the other direction, 211,199, both resulting numbers are prime numbers.

Generating a Large Prime with the Number 206

The number 206 enables us to reach a prime number in a rather curious fashion:

$206^0 + 206^1 + 206^2 + 206^3 + 206^4 + 206^5 + 206^6 = 76{,}792{,}124{,}280{,}187$, which is a prime number.

Generating a Series of Primes Where the Difference Is the Number 210

A curious aspect of the number 210 is that a series of prime numbers exists where the common difference is 210, as shown here: 47, 257, 467, 677, 887, 1097, and 1307.

Surprises of Primes and Palindromes from the Number 229

When the number 229 is added to its reversal 922, the resulting number, 1,151, is the 190th prime number.

There are times when strange manipulations create prime numbers. Suppose we replace each digit of the number 229 with its square to get the number 4,481, which is the 608th prime number. Now we will replace each digit with its cube to get the number 88,729, which is the 8,593rd prime number. If we take the difference of these two numbers, $88{,}729 - 4{,}481 = 84{,}248$, which, although not a prime number, is an attractive palindromic number. When we add these two numbers to 229, we get $4481 + 88729 + 229 = 93{,}439$, another unexpected palindromic. Surprises never end!

Curiosities of Reversals with the Number 241

The number 241 is a prime number and the twin prime partner of 239. When 241 is added to its reversal 142, the result is a palindromic number, 383. From another point of view, the number 241 can be expressed as the sum of a number and its reversal: $241 = 170 + 071$.

The number 241 can be set equal to a configuration using only the digit 2, as follows:

$$\frac{22^2 - 2}{2} = 241.$$

Uniqueness of the Number 256

The number 256 is equal to the 8th power of 2, namely, 2^8.

The famous Hungarian mathematician Paul Erdös (1913–1996) conjectured that the highest power of 2 that can be expressed as a sum of distinct powers of 3 is the number 256. In other words, $256 = 3^5 + 3^2 + 3^1 + 3^0$.

Strange Patterns of the Number 271

The number 271 is a twin prime number with 269, and strangely enough, 271 is a factor of the repdigit number 11,111, since $271 \times 41 = 11{,}111$.

The reversals of the second and third powers of the number 271 are prime numbers. Consider $271^2 = 73{,}441$, and then its reversal 14,437 is a prime number. Considering the number $271^3 = 19{,}902{,}511$, we find that its reversal, 11,250,991, is also a prime number.

The reciprocal of 271 also has an unusual five-place repetition as we can see here:

$$\frac{1}{271} = 0.00369\underline{00369}00369...$$

Three Ways Consecutive Primes Generate the Composite Number 287

Although the number 287 is not a prime number, it can be obtained from three different sums of consecutive prime numbers as follows:

- $287 = 89 + 97 + 101$;
- $287 = 47 + 53 + 59 + 61 + 67$;
- $287 = 17 + 19 + 23 + 29 + 31 + 37 + 41 + 43 + 47$.

The Kaprekar Number 297

The number 297 is the fifth Kaprekar number. To demonstrate this, we take square of 297 to get $297^2 = 88{,}209$, split it into two parts, and, strangely enough, the sum of the two numbers of the split results in the original number: $88 + 209 = 297$. Such a number is called a *Kaprekar number*, and we once again credit the Indian mathematician Dattaraya Ramchandra Kaprekar who discovered such unusual number relationships. (For more about Kaprekar numbers see page 94.)

If we take a cyclic permutation of the number 297, say 729, raise it to the second power, $729^2 = 531{,}441$, and then split the number to get the numbers 531 and 441, amazingly, we find that their sum is 972, which is the number we started with.

The Strange Number 376

The number 376 shares a property with the number 625 in that its powers all end in the same three digits from which they started. We can see that in the list in Figure 1-15:

376^k		
376^1	=	**376**
376^2	=	141,**376**
376^3	=	53,157,**376**
376^4	=	19,987,173,**376**
376^5	=	7,515,177,189,**376**
376^6	=	2,825,706,623,205,**376**
376^7	=	1,062,465,690,325,221,**376**
376^8	=	399,487,099,562,283,237,**376**
376^9	=	150,207,149,435,418,497,253,**376**
376^{10}	=	56,477,888,187,717,354,967,269,**376**
376^{11}	=	21,235,685,958,581,725,467,693,285,**376**
376^{12}	=	7,984,617,920,426,728,775,852,675,301,**376**

Figure 1-15

The Perfect Number 496

The number 496 is a perfect number since it is equal to the sum of its proper divisors, namely, $1 + 2 + 4 + 8 + 16 + 31 + 62 + 124 + 248 = 496$.

The Strange Number 625

When the number 625 is squared, the result is 390,625. When the number 625 is cubed, we get $625^3 = 244{,}140{,}625$.

Taking this a step further, $625^4 = 152{,}587{,}890{,}625$, and $625^5 = 95{,}367{,}431{,}640{,}625$. What should be noticed here is that every power of 625 ends in the same three digits, namely, 625. Figure 1-16 exhibits the first few cases of this phenomenon.

625^k		
625^1	=	**625**
625^2	=	390,**625**
625^3	=	244,140,**625**
625^4	=	152,587,890,**625**
625^5	=	95,367,431,640,**625**
625^6	=	59,604,644,775,390,**625**
625^7	=	37,252,902,984,619,140,**625**
625^8	=	23,283,064,365,386,962,890,**625**
625^9	=	14,551,915,228,366,851,806,640,**625**
625^{10}	=	9,094,947,017,729,282,379,150,390,**625**
625^{11}	=	5,684,341,886,080,801,486,968,994,140,**625**

Figure 1-16

The Delectable Number 666

Here are a few delectable number relationships that lead to 666:

$$666 = 1^6 - 2^6 + 3^6$$
$$666 = (6+6+6) + (6^3 + 6^3 + 6^3)$$
$$666 = (6^4 - 6^4 + 6^4) - (6^3 + 6^3 + 6^3) + (6+6+6)$$
$$666 = 5^3 + 6^3 + 7^3 - (6+6+6)$$
$$666 = 2^1 \cdot 3^2 + 2^3 \cdot 3^4.$$

We can even generate 666 by representing each of its three digits in terms of 1, 2, and 3:

$$6 = 1+2+3$$
$$6 = 1 \cdot 2 \cdot 3$$
$$6 = \sqrt{1^3 + 2^3 + 3^3}$$

Therefore, $666 = (100)(1+2+3) + (10)(1 \cdot 2 \cdot 3) + \left(\sqrt{1^3 + 2^3 + 3^3}\right).$

Curious Rearrangements of the Number 691

The number 691 is a prime number but none of its digits are prime numbers. When the number 691 is reversed to get 196, a perfect square number results: $196 = 14^2$. When the number 691 is flipped over to get 961, we have another perfect square: $961 = 31^2$. We reverse 961 to get 169, another perfect square number: $169 = 13^2$.

The Flipped Number 836

The number 836, when it is squared, results in a rather unusual number, namely, 698,896. The number 698,896 is not only a palindrome but also can be flipped upside down to become the palindrome

968,869 = 3 × 32,323, and thus another palindrome appears. This time, the palindrome 32,323 is also a prime number.

Limited Extensions to Produce Primes with the Number 701

The number 701 is the lead number of the following prime numbers: 701, 7001, 70001, and 700001. This cannot be extended by inserting five 0s, since 7000001 = 197 × 35,533 and is not a prime. The prime number $701 = 1^0 + 2^1 + 3^2 + 4^3 + 5^4$.

Nice Pattern for the Prime Number 733

The number 733 is the 130th prime number.

The number $733 = 1^7 + 2^6 + 3^5 + 4^4 + 5^3 + 6^2 + 7^1 + 8^0$. Notice the pattern of the exponents!

A Curious Pattern with the Numbers 788–793

The consecutive numbers 788, 789, 790, 791, 792, and 793 are divisible by the consecutive prime numbers 2, 3, 5, 7, 11, and 13, respectively. Rather curious!

Unique Representation for the Number 823

The number 823 is a prime number with a unique feature:

$$823 = \frac{12{,}345}{1+2+3+4+5}.$$

Weird Creation of Primes from the Number 907

Beginning with the number 907, which is a prime number, a series of primes can be constructed as follows: 907, 9907, 99907, 999907, 9999907, are each prime numbers.

A Palindromic Arrangement with the Palindromic Number 919

A cute relationship offered by the number 919 is the following:

$$919 = 91 + 81 + 71 + 61 + 51 + 41 + 31 + 21 + 11 + 1 + 11 + 21 + 31 + 41 + 51 + 61 + 71 + 81 + 91.$$

Surprising Sums and Products of the Digits of the Number 981

The digits of the number 981 can be partitioned into five groups of three numbers each, where each group has the same sum and the same product, as shown here:

- $981 = 6 + 480 + 495$, and $6 \times 480 \times 496 = 1{,}425{,}600$;
- $981 = 11 + 160 + 810$, and $11 \times 160 \times 810 = 1{,}425{,}600$;
- $981 = 12 + 144 + 825$, and $12 \times 144 \times 825 = 1{,}425{,}600$;
- $981 = 20 + 81 + 880$, and $20 \times 81 \times 880 = 1{,}425{,}600$;
- $981 = 33 + 48 + 900$, and $33 \times 48 \times 900 = 1{,}425{,}600$.

This is just one of the things that mathematics allows a creative mind to pursue.

The Magic of the Number 1089

The number 1089 has a number of oddities attached to it. One characteristic of this number can be seen by finding its reciprocal:

$$\frac{1}{1089} = .\overline{0009182736455463728191}.$$

With the exception of the first three zeros and the last 1, we have a palindromic number 918,273,645,546,372,819, since it reads the same in both directions.

Furthermore, if we multiply 1089 by 5, we also get a palindromic number 5445, and if we multiply 1089 by 9, we get 9801, the reverse of the original number. By the way, the only other number of four or fewer digits whose reverse is a multiple of the original number is 2178, since $2{,}178 \times 4 = 8{,}712$.

The number $1089 = 33^2 = 65^2 - 56^2$. This is the only two-digit illustration of this pattern.

Let us now multiply 9 by some modifications of 1089, say 10989, 109989, 1099989, and 10999989, and then marvel at the results:

$$10989 \times 9 = 98901$$
$$109989 \times 9 = 989901$$
$$1099989 \times 9 = 9899901$$
$$10999989 \times 9 = 98999901$$

and so on.

Returning to the number 1089, we find embedded in it a very entertaining oddity that actually made this number 1089 famous. Suppose you select any three-digit number whose units digit and hundreds digits are not the same and then reverse that number. Now subtract the two numbers you have (obviously, the larger minus the smaller). Once again reverse the digits of this arrived at difference, and add this new number to the difference. The result will always be 1089.

To see that this works, we will choose any randomly selected three-digit number, say 732. We now subtract $732 - 237 = 495$. Reversing the digits of 495 we get 594, and now we add these last two numbers: $495 + 594 = 1089$. Yes, this will hold true for all such three-digit numbers—amazing! This is a cute little "trick" that can be justified with simple algebra.

For the motivated reader we provide the justification of this technique using elementary algebra. We shall represent the arbitrarily selected three-digit number $\overline{htu}$ as $100h + 10t + u$, where h represents the hundreds digit, t represents the tens digit, and u represents the units digit. The number with the digits reversed is then: $100u + 10t + h$. We will let $h > u$, which would be the case either in the number you selected or its reverse. In the subtraction, $u - h < 0$; therefore, take 1 from the tens place (of the minuend) making the units place $10 + u$. Since the tens digits of the two numbers to be subtracted are equal, and 1 was taken from the tens digit of the minuend, then the value of this digit is $10(t - 1)$. The hundreds digit of the minuend is $h - 1$, because 1 was taken away to enable subtraction in the tens place, making the value of the tens digit $10(t-1) + 100 = 10(t+9)$.

When we do the first subtraction, we get:

$$\begin{array}{lll} 100(h-1) & +10(t+9)+(u+10) & \\ 100u & +10t & +h \\ \hline 100(h-u-1) & +10(9) & +u-h+10 \end{array}$$

Reversing the digits of this difference $100(h - u - 1) + 10 \times 9 + (u - h + 10)$ gives us:

$100(u - h + 10) + 10 \times 9 + (h - u - 1)$. Adding these last two expressions gives us:

$100(h - u - 1) + 10 \times 9 + (u - h + 10) + 100(u - h + 10) + 10 \times 9 + (h - u - 1) = 1000 + 90 - 1 = \mathbf{1089}$.

This algebraic justification enables us to inspect the general case of this arithmetic process, allowing us to guarantee that this process holds true for all numbers.

Interesting Property of the Number 1111

The number 1111 is the sum of the digits of the first 100 prime numbers. The motivated reader may wish to verify it with the first 100 primes shown in Figure 1-17:

2	3	5	7	11	13	17	19	23	29
31	37	41	43	47	53	59	61	67	71
73	79	83	89	97	101	103	107	109	113
127	131	137	139	149	151	157	163	167	173
179	181	191	193	197	199	211	223	227	229
233	239	241	251	257	263	269	271	277	281
283	293	307	311	313	317	331	337	347	349
353	359	367	373	379	383	389	397	401	409
419	421	431	433	439	443	449	457	461	463
467	479	487	491	499	503	509	521	523	541

Figure 1-17

Interesting Number 1233

The number 1233 is one of those unusual numbers that can be broken into the following pattern: $1233 = 12^2 + 33^2$.

Unusual Way to Generate Primes with the Prime Number 1361

Even though the number 1361 is a prime number it allows the creation of other prime numbers in a rather unusual way. By splitting up the 13 and the 61 and placing a **63** between them, one gets the number 13**63**61, which is a prime number. This can be repeated by inserting two numbers 63 between the 13 and the 61 to get the number 13**6363**61, which is also a prime number. This can be repeated twice more to get two additional prime numbers, namely: 13**636363**61 and 13**63636363**61. However, when another 63 is inserted, the result is a composite number 13**6363636363**61 = $7 \times 1321 \times 5693 \times 10661$. Mathematics is curious!

Powers of 10 Generating Primes from the Number 1399

The number 1399 is a prime number that can generate further primes by adding consecutive powers of 10 to it. Therefore, the following numbers are all prime numbers:

$$1399 + 10 = 1409$$
$$1399 + 100 = 1499$$
$$1399 + 1000 = 2399$$
$$1399 + 10{,}000 = 11{,}399$$
$$1399 + 100{,}000 = 101{,}399.$$

Unexpected Reversals from the Number 1453

The number 1453 is the largest prime number that can be represented in two different ways as a sum of a three-digit number and its reversal: $1453 = 479 + 974$, and $1453 = 677 + 776$.

The Year of the Number 2025

We have now come to the year this book was written, 2025. At first glance one might look at this number 2025, and simply correctly conclude that it must be a multiple of 5 since the last digit is a 5, as well as a multiple of 25 since the last two digits are 25. But there are usual aspects to any number, even if its appearance is as commonplace as the number 2025. First, we notice that the number 2025 is a perfect square, that is, $45^2 = 2025$. At the outset this does not seem strange. However, only people over 90 years old have experienced another year which was a perfect square, namely, $1936 = 44^2$. The next year that will be a perfect square is $2116 = 46^2$, which many of us are not likely to see. The fact that the year 2025 is the only perfect square year that many of us will experience is only a part of the attraction of this number. For example, if we split the number and add the two parts we find 20 + 25 = 45, which is the square root of 2025. Mathematicians are always searching for peculiarities involving numbers. We also notice that $45 = 9 \cdot 5$, and $9^2 \cdot 5^2 = 81 \cdot 25 = 2025$. What's more, 2025 can be also expressed as a sum of three squares: $5^2 + 20^2 + 40^2 = 2025$.

A relationship of two sets of numbers discovered by the Greek mathematician Nicomachus (60 CE – 120 CE) can be applied to 2025. In general terms, Nicomachus's theorem states that the sum of cubes of first n natural numbers is equal to the square of the sum of these n natural numbers. That is, $1^3 + 2^3 + 3^3 + 4^3 + 5^3 + 6^3 + 7^3 + 8^3 + \cdots + n^3 = (1+2+3+4+5+6+7+8+\cdots+n)^2$. Amazingly, when applied to 2025, we find that $1^3 + 2^3 + 3^3 + 4^3 + 5^3 + 6^3 + 7^3 + 8^3 + 9^3 = 2025$, and also that $(1+2+3+4+5+6+7+8+9)^2 = 45^2 = 2025$.

The number 2025 is also divisible by the sum of its digits, namely,

$$\frac{2025}{2+0+2+5} = 225.$$

Furthermore, the number 2025 is equal to the sum of the first 45 odd numbers and be expressed as $1+3+5+7+\cdots+87+89 = 2025$. As if this is not enough to convince that numbers are spectacular, consider the multiplication table for 1 to 9 shown in Figure 1-18.

X	1	2	3	4	5	6	7	8	9
1	1	2	3	4	5	6	7	8	9
2	2	4	6	8	10	12	14	16	18
3	3	6	9	12	15	18	21	24	27
4	4	8	12	16	20	24	28	32	36
5	5	10	15	20	25	30	35	40	45
6	6	12	18	24	30	36	42	48	54
7	7	14	21	28	35	42	49	56	63
8	8	16	24	32	40	48	56	64	72
9	9	18	27	36	45	54	63	72	81

Figure 1-18

Quite unexpectedly, we find that the sum of the products from 1 to 81 is equal to 2025.

There is another uniqueness about the number $45^2 = 2025$ that is little bit contrived. Consider a set of numbers where the actual number determines the amount of entries that number will have into the set. That is, the number 1 will appear once, the number 2 will appear twice, the number 3 will appear three times, and so on until we reach the number 45, which will appear 45 times. The set will look like this: $\{1,2,2,3,3,3,4,4,4,4,5,5,5,5,5,....45,45,45,45...\}$. Unexpectedly, the number of digits in this set is $45^2=2025$. No other number in our number system greater than 1 has this property.

If you want to stretch it a bit, you can take a date in the year 2025 and find that it is a square date. September 27, 2025, is written in the United States as 09/27/2025, and written as a number it would be $9{,}272{,}025 = 3045^2$. The date would be written in Europe as 27/09/2025, which when written as a number is $27{,}092{,}025=5205^2$. You might want to consider this date as a mathematically special date in the year 2025.

The Magic of the Number 2178

The number 2178 has a magical trait to it, for when it is multiplied by 4, its reversal appears: $2178 \times 4 = 8712$. If this is not enough of an amazement, then consider placing a 9 in the middle of the number 2178 to get 21978. Once again, we repeat the process: $21978 \times 4 = 87912$, which is its reversal. Surprisingly, this can be continued by placing 99 in the middle of 2178 to get 219978, and when multiplied by 4, we get $219978 \times 4 = 879912$, once again the reversal of the number with which we started. This can then be continued by placing 999 in the middle of 2178 to get its reversal by multiplying by 4: $2199978 \times 4 = 8799912$. This pattern continues.

Amazing Trick with the Number 2418

We will use the number 2418 to demonstrate a clever little arithmetic trick. This trick only holds for numbers where two adjacent digits do not have a sum exceeding 9. When we multiply $2418 \times 11 = 26598$, and then multiply its reversal by 11, we get $8142 \times 11 = 89562$, which is the reverse of the previous product we obtained.

Uniqueness of the Number 3435

The number 3435 seems to be the only multi-digit number that has the following property:

$$3435 = 3^3 + 4^4 + 3^3 + 5^5.$$

The Magic Number 6174

The number 6174 is known as the *Kaprekar constant*, named for the famous Indian recreational mathematician Dattatreya Ramchandra Kaprekar. This constant arises when one takes any four-digit number and rearranges the digits to create the largest and the smallest number from these digits and then subtracts these two newly formed numbers. Continuing this process will always eventually result in the number 6174. Upon reaching the number 6174, if we continue the process of creating the largest and the smallest number and then taking their difference (7,641 – 1,467 = 6,174), we notice that we get back to 6174. The number 6174 is, therefore called the *Karprekar constant*. To demonstrate this with an example, we will carry out this process with the number 2303 as follows:

- The largest number formed with these digits is: 3320.
- The smallest number formed with these digits is: 0233.
- The difference is: 3087.
- The largest number formed with these digits is: 8730.
- The smallest number formed with these digits is: 0378.
- The difference is: 8352.
- The largest number formed with these digits is: 8532.
- The smallest number formed with these digits is: 2358.
- The difference is: 6174.
- The largest number formed with these digits is: 7641.
- The smallest number formed with these digits is: 1467.
- The difference is: 6174.

And so the loop is formed, since you continue to get the number 6174. Remember, all this began with an arbitrarily selected four-digit number. Processing a four-digit number in this way will always end with the number 6174, which then gets you into an endless loop, that is, continuously getting back to 6174.

Another curious property of 6174 is that it is divisible by the sum of its digits:

$$\frac{6174}{6+1+7+4}=\frac{6174}{18}=343.$$

Astonishing Property of the Number 6666

The square of the number 6666 has a peculiar characteristic, where $6666^2 = 44435556$, which when separated into two parts and added together results in 4443 + 5556 = 9999.

This can also be done with the number 3333^2 and 7777^2. Verifying this is left as an entertainment for the reader.

The Perfect Number 8128

After the first three perfect numbers, 6, 28, and 496, the number 8128 is the fourth perfect number, as it is the sum of its proper divisors:

$$1 + 2 + 4 + 8 + 16 + 32 + 64 + 127 + 254 + 508 + 1016 + 2032 + 4064 = 8128.$$

Unusual Number 9474

The number 9474 is one of three four-digit numbers that can be expressed as a sum of the fourth power of its digits, as we show here: $9474 = 9^4 + 4^4 + 7^4 + 4^4$.

The Unique Number 9999

The number 9999 has some unusual characteristics. We can take the square of 9999 to get $9999^2 = 99980001$, then split that into two parts, and find the sum to be $9998 + 0001 = 9999$, which is the number we began with.

Now consider the cube of the number 9999, which is $9999^3 = 999700029999$. When we take the sum of those three equal-length parts, we get $9997 + 0002 + 9999 = 19998$, and $19998 + 2 = 20000$. Unfortunately, this cannot be extended to the fourth power.

The Amazing Number 142,857

The number 142,857 is particularly attractive for its various unusual aspects. When the number 142,857 is multiplied by the numbers from 1 to 6, the result shows a cyclic permutation of the digits of the original number, as we show below:

$$\begin{aligned}
142{,}857 \times 1 &= 142{,}857 \\
142{,}857 \times 2 &= 285{,}714 \\
142{,}857 \times 3 &= 428{,}571 \\
142{,}857 \times 4 &= 571{,}428 \\
142{,}857 \times 5 &= 714{,}285 \\
142{,}857 \times 6 &= 857{,}142.
\end{aligned}$$

When this number 142,857 is multiplied by 7, the product is 999,999. Multiplying the number 142,857 by larger numbers produces some other interesting results that will be worth the readers' efforts to explore.

The number 142,857 can be partitioned into two parts and added: $142 + 857 = 999$. It should be noted that the number 999 is also a factor of 142,857, since $142{,}857 \div 999 = 143$.

The nines can reappear by adding the pairs of digits of our number 142,857 as follows:

$$14 + 28 + 57 = 99.$$

This can be taken a giant step further if we take our number 142,857 and multiply it by any large number, say 243, and then split the number and marvel over the result. First, $142{,}857 \times 243 = 34{,}714{,}251$. Then partitioning the number in thirds from right to left: $34 + 714 + 251 = 999$. For a further example, $142{,}857 \times 3636 = 519{,}428{,}052$, then by partitioning and adding we get $519 + 428 + 052 = 999$.

We know that the number 9 is a factor of 142,857, since the sum of the digits $1 + 4 + 2 + 8 + 5 + 7 = 27$, which is a multiple of 9. Therefore, the original number $142{,}857 \div 9 = 15{,}873$.

When we square the number 142,857 and split it another interesting result emerges:

$142{,}857^2 = 20{,}408{,}122{,}449$. By splitting this number and adding the two parts:

$20{,}408 + 122{,}449 = 142{,}857$, we get back to the original number.

The reciprocal of 142,857 also produces a surprise: $\frac{1}{142857} = 0.000007$. The appearance of the 7 could let us expose an extra surprise within the reciprocal of 7: $\frac{1}{7} = 0.\overline{142857}$. Lo and behold, our original number 142,857 reappears in the repeating decimal.

Furthermore, the number 142,857 is divisible by both 11 and 111, since $142857 = 111 \times 1287 = 11 \times 12987$. Thus, the number 142,857 stores many unusual characteristics.

The Unexpected Feature of the Number 175,560

The number 175,560 can be expressed as the product of consecutive numbers in two different ways: $175{,}560 = 55 \times 56 \times 57 = 19 \times 20 \times 21 \times 22$.

The Amazing Number 193,939

We now present a truly remarkable prime number: 193,939. Were we to write it in reverse, we would get 939,391, which is also a prime number. By considering other permutations of this number, we get the prime numbers shown here:

193,939
939,391
393,919
939,193
391,939
919,393

The Incredible Number 490,689

The number 490,689 can be broken up into the sum of three cubes in two different ways, leading to a further surprise. $490{,}689 = 4^3 + 60^3 + 65^3 = 8^3 + 25^3 + 78^3$. Now comes the surprise. By using the same base numbers, we find an incredible coincidence of equality: $4 \times 60 \times 65 = 8 \times 25 \times 78$.

The Surprising Number 548,834

The number 548,834 has one peculiar aspect to it, namely, the number can be expressed as the sum of the sixth powers of its digits: $548{,}834 = 5^6 + 4^6 + 8^6 + 8^6 + 3^6 + 4^6$.

The Unexpected Primes Embedded in the Number 739,397

The number 739,397 has a most unusual characteristic. If we remove one digit at a time from left to right, we always create a new prime number, as we can show here with the following prime numbers: 739,397, then 39,397, then 9,397, then 397, then 97, and finally 7. This can also be done in the opposite direction from right to left to once again generate a list of prime numbers: 739,397, then 73,939, then 7,393, then 739, then 73, and finally 7.

The Amazing Number 438,579,088

The number 438,579,088 has a rather unusual characteristic, namely, that the sum of the digits taken to the power indicated by the digit yields the original number 438,579,088.

Admire this relationship:

$$438{,}579{,}088 + 1 = 4^4 + 3^3 + 8^8 + 5^5 + 7^7 + 9^9 + 0^0 + 8^8 + 8^8.$$

The Unique Digit Number 9,876,543,210

The number 9,876,543,210 is a number where all digits are represented exactly once. When we subtract from this number its reversal, 0,123,456,789, the answer is 9,753,086,421. The unusual result is that each of these three numbers is comprised of all the digits used exactly once.

Key Number Properties

Prime Numbers

Prime numbers are numbers that have exactly two divisors. The smallest prime number is 2, as its divisors are 1 and 2. Another example of a prime number is the number 7, since its only factors are 1 and 7. Nonprime numbers greater than 1 are called *composite numbers*, as they have more than two factors. Figure 1-19 provides a table of prime numbers less than 1000.

Perfect Numbers

Although we have covered perfect numbers at various times in the preceding chapter, such the number 6 and the number 28, you may be wondering now what makes a number perfect? Mathematicians define a *perfect number* to be a number equal to the sum of its factors (excluding the number itself). For example, the smallest perfect number is 6, because the sum of its factors (its proper positive divisors) is $1 + 2 + 3 = 6$. The next perfect number is 28, since the sum of its factors is $1 + 2 + 7 + 4 + 14 = 28$. Perfect numbers have fascinated mathematicians for centuries. The ancient Greeks knew of the first four perfect numbers, and Euclid (ca. 300 BCE) even established a formula for generating perfect numbers. These first two perfect numbers were regarded by ancient Biblical scholars as perfect because the

Prime Numbers
1 – 1000

	2	3	5	7	11	13	17	19	23
29	31	37	41	43	47	53	59	61	67
71	73	79	83	89	97	101	103	107	109
113	127	131	137	139	149	151	157	163	167
173	179	181	191	193	197	199	211	223	227
229	233	239	241	251	257	263	269	271	277
281	283	293	307	311	313	317	331	337	347
349	353	359	367	373	379	383	389	397	401
409	419	421	431	433	439	443	449	457	461
463	467	479	487	491	499	503	509	521	523
541	547	557	563	569	571	577	587	593	599
601	607	613	617	619	631	641	643	647	653
659	661	673	677	683	691	701	709	719	727
733	739	743	751	757	761	769	773	787	797
809	811	821	823	827	829	839	853	857	859
863	877	881	883	887	907	911	919	929	937
941	947	953	967	971	977	983	991	997	

Figure 1-19

Biblical creation was accomplished in 6 days and the lunar month has 28 days. The discovery of the next two perfect numbers, 496 and 8128, is attributed to the Greek mathematician Nicomachus (60–120 CE). Yet, it was not until the eighteenth century that the Swiss mathematician Leonhard Euler (1707–1783) proved that Euclid's formula would generate all even perfect numbers.

Let us now consider Euclid's generalized formula for generating additional perfect numbers. The formula for generating even perfect numbers is $2^{n-1}(2^n - 1)$, with the condition that $2^n - 1$ equals a prime number; this prime number is called a Mersenne prime number, and represented as M_p.[2] Crucially, if n is a composite number, then $2^n - 1$

[2] $M_p = 2^p - 1$ and p is prime.

will surely be composite, thereby making a composite value of n incapable of generating a perfect number from Euclid's formula. We can demonstrate this with some very elementary algebra, as we will show here.

Suppose n is an even composite number, say $2x$. The expression $2^n - 1$ then becomes the difference of two squares, which is always factorable, and is, therefore, a composite number: $2^{2x} - 1 = (2^x - 1)(2^x + 1)$. If n is an odd composite number, as in the expression $2^{pq} - 1$, it becomes a factorable term as

$$2^{pq} - 1 = \left(2^p\right)^q - 1 = \left(2^p - 1\right)\left(\left(2^p\right)^{q-1} + \left(2^p\right)^{q-2} + \left(2^p\right)^{q-3} + \cdots + \left(2^2\right) + 1\right).$$

This is not to say that whenever n is prime that $2^n - 1$ will also be prime. For example, when $n = 11$, we have $2^{11} - 1 = 2048 - 1 = 2047 = 23 \times 89$, and is, therefore, not prime. Hence, we have to be careful to make sure that n is a prime number and that it also generates $2^n - 1$ to be a prime.

Mathematicians are always fascinated with perfect numbers, and consequently are always in search for further members of this set. As of the publication date, there are 52 known perfect numbers. The table in Figure 1-20 shows the first ten perfect numbers.

Rank	n	Perfect Number $\left(2^{n-1}\right)\left(2^n - 1\right)$	Digits	Year Discovered
1	2	6	1	Greeks
2	3	28	2	Greeks
3	5	496	3	Greeks
4	7	8128	4	Greeks
5	13	33550336	8	1456
6	17	8589869056	10	1588
7	19	137438691328	12	1588
8	31	2305843008139952128	19	1772
9	61	2658455991569831744654692615953842176	37	1883
10	89	191561942608236107294793378084303638130997321548169216	54	1911

Figure 1-20

There are lots of curious characteristics of perfect numbers. For example, they all end in either 28 or 6, and that preceded by an odd digit. Mathematicians have been in search for odd perfect numbers, and as we can see our list consists only of even perfect numbers. To date, they say, with confidence, that there are no odd perfect numbers less than 10^{1500}.

There are many unusual characteristics of perfect numbers beyond those that define them—namely, that the sum of their divisors equals the number itself. Here are some curious characteristics of perfect numbers of the form $2^{n-1}(2^n - 1)$.

First, they are the sum of the first consecutive natural numbers, as shown below:

$$6 = 1 + 2 + 3$$
$$28 = 1 + 2 + 3 + 4 + 5 + 6 + 7$$
$$496 = 1 + 2 + 3 + 4 + 5 + 6 + 7 + 8 + 9 + 10 + 11 + 12 + 13 + 14 + \cdots + 29 + 30 + 31$$
$$8128 = 1 + 2 + 3 + 4 + 5 + 6 + 7 + 8 + 9 + 10 + 11 + 12 + 13 + 14 + \cdots + 125 + 126 + 127$$
$$33550336 = 1 + 2 + 3 + 4 + 5 + 6 + 7 + 8 + 9 + 10 + 11 + 12 + 13 + \cdots + 8189 + 8190 + 8191,$$

and so on.

If this isn't enough, we also notice, with the exception of the first perfect number 6, they are equal to the sum of consecutive odd cubes, as shown below:

$$28 = 1^3 + 3^3$$
$$496 = 1^3 + 3^3 + 5^3 + 7^3$$
$$8{,}128 = 1^3 + 3^3 + 5^3 + 7^3 + 9^3 + 11^3 + 13^3 + 15^3$$
$$33{,}550{,}336 = 1^3 + 3^3 + 5^3 + \cdots + 123^3 + 125^3 + 127^3.$$

How, you may wonder, can it be that suddenly perfect numbers are equal to the sum of the cubes of a sequence of odd numbers? This can be easily justified through some elementary algebra. We recall that the perfect number must take the form of $2^{n-1}(2^n - 1)$, where $2^n - 1$ is a prime number. We will take a moment here to show how we can justify that each perfect number is the sum of the first 2^k odd numbers, where $k = \frac{1}{2}(n-1)$, except for $n = 2$. We should recall that the following are true:

$$S_1 = 1+2+3+4+\cdots+q = \frac{q(q+1)}{2}$$

$$S_2 = 1^2+2^2+3^2+4^2+\cdots+q^2 = \frac{q(q+1)(2q+1)}{6}$$

$$S_3 = 1^3+2^3+3^3+4^3+\cdots+q^3 = \frac{q^2(q+1)^2}{4} = S_1^2.$$

Let us now look at the sum of the cubes of the odd numbers. We can write that as follows:

$$S = 1^3+3^3+5^3+7^3+\cdots+(2q-1)^3 = \sum_{i=1}^{q}(2q-1)^3.$$

With some algebraic manipulation (working backwards), we can show that this is equal to the following:

$$\begin{aligned} S &= \sum_{i=1}^{q}(2q-1)^3 - \sum_{i=1}^{q}(8q^3 \quad 12q^2+6q-1) \\ &= 8\cdot\frac{q^2(q+1)^2}{4} - 12\cdot\frac{q(q+1)(2q+1)}{6} + 6\cdot\frac{q(q+1)}{2} \quad q = q^2(2q^2-1). \end{aligned}$$

Notice how we inserted the first three formulas, which we generated above, into this last equation. If we now let $q = 2^k$, then $S = 2^{2k}(2^{2k+1} - 1)$. This would generate the perfect numbers, $2^{n-1}(2^n - 1)$, when $n = 2k + 1$, which are the odd numbers. This shows how we go from the general formula for perfect numbers to the sum of the cubes of the odd numbers. Although this algebra may be a little bit tough, we provided it here to show how we can justify some of these mathematical curiosities.

All perfect numbers must have an even number of divisors. If we take the reciprocals of the divisors of any perfect number (now including the number itself), their sum will always be equal to 2. This can be seen from the first few perfect numbers as follows:

$$\frac{1}{1}+\frac{1}{2}+\frac{1}{3}+\frac{1}{6}=2$$

$$\frac{1}{1}+\frac{1}{2}+\frac{1}{4}+\frac{1}{7}+\frac{1}{14}+\frac{1}{28}=2$$

$$\frac{1}{1}+\frac{1}{2}+\frac{1}{4}+\frac{1}{8}+\frac{1}{16}+\frac{1}{31}+\frac{1}{62}+\frac{1}{124}+\frac{1}{248}+\frac{1}{496}=2.$$

Although we already discussed many of these properties in Chapter 1, we are offering a deeper understanding of the concepts involved in perfect numbers for the interested reader.

Palindromic Numbers

Palindromic numbers are those that read the same in both directions. The first four powers of 11 are palindromic numbers:

$$11^1=11$$

$$11^2=121$$

$$11^3=1331$$

$$11^4=14641.$$

As we have experienced, a palindromic number can either be a prime number or a composite number. For example, 151 is a prime palindrome and 171 is a composite palindrome. Yet, with the exception of 11, a palindromic prime must have an odd number of digits.

A palindromic number can be generated from any given number. All we need to do is to continually add a number to its reversal, which is the number written in the reverse order of digits, until a palindrome results. We can even reach a palindrome with a single addition. For example, starting with the number 23, we have 23 + 32 = 55, a palindrome. Or it might take two steps to reach a palindrome, such as with the starting number 75: 75 + 57 = 132, 132 + 231 = 363, a palindrome.

Or it might take three steps, such when we start with 86:

86 + 68 = 154, 154 + 451 = 605, 605 + 506 = 1111, a palindrome.

The starting number 97 will require 6 steps to reach a palindrome, while the number 98 will require 24 steps to reach the palindromic number 8,813,200,023,188.

Now we come to the number 196. Many attempts have been made to use this rule of reverse-and-add to eventually reach a palindromic number from 196. So far, this has never been accomplished. As a matter of fact, the number 196 is the smallest number that (to the best of our current knowledge) does not generate a palindromic number through the reverse-and-add rule. The next few numbers that do not generate a palindromic number with this procedure are: 295, 394, 493, 592, 689, 691, 788, 790, 879, 887, ...

There are some lovely patterns when dealing with palindromic numbers. For example, numbers that yield palindromic cubes are palindromic themselves:

$$
\begin{aligned}
11^3 &= 1{,}331\\
101^3 &= 1{,}030{,}301\\
1001^3 &= 1{,}003{,}003{,}001\\
10001^3 &= 1{,}000{,}300{,}030{,}001\\
100001^3 &= 1{,}000{,}030{,}000{,}300{,}001
\end{aligned}
$$

... and this continues on.

Triangular Numbers

A number is triangular if that number of points can be arranged into an equilateral triangle, as shown in Figure 1-21. The first few triangular numbers are: 0, 1, 3, 6, 10, 15, 21, 28, 36, 45, 55, 66, 78, 91, 105, 120, 136, 153, 171, 190, 210, 231, 253, 276, 300, 325, 351, 378, 406, 435, 465, 496, 528, 561, 595, 630, 666.... The number 3 has a distinct role to the famous German mathematician Carl Friedrich Gauss (1777–1855), who showed that every integer is the sum of at most three triangular numbers. For example, 39 = 28 + 10 + 1, 18 = 15 + 3, 51 = 45 + 6, 95 = 91 + 3 + 1, etc.

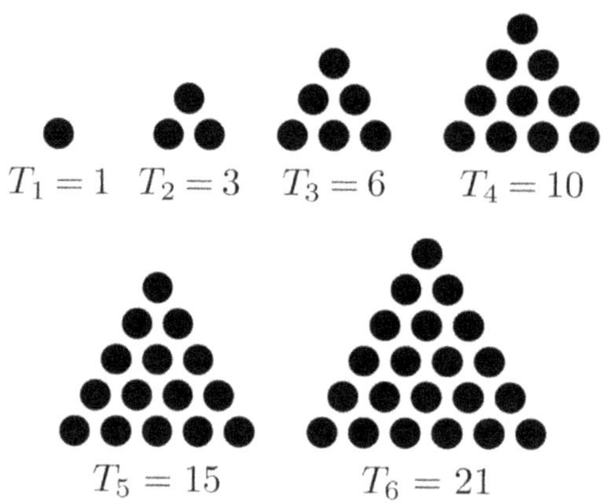

Figure 1-21

Kaprekar Numbers

If you take the square of a number, split this result in two parts, add the two parts together, and the sum is the original number, you have found a Kaprekar number. The number 297 is a Kaprekar number. To demonstrate this characteristic, we take the square of 297, which is $297^2 = 88{,}209$, split it into two parts, add the two numbers resulting from the split, and what results is the original number: 88 + 209 = 297. Such a number is called a *Kaprekar number*, named after the Indian recreational mathematician Dattaraya Ramchandra Kaprekar (1905–1986), who discovered such numbers and a variety of other number relationships that bear his name. Another example of a Kaprekar number is the number 703, which is a Kaprekar number since $703^2 = 494{,}209$, and $494 + 209 = 703$. Some other Kaprekar numbers are shown in the table in Figure 1-22.

Kaprekar Numbers		**Square of the Numbers**				**Decomposition of the Numbers**
1		1^2	=	1		1 = 1
9		9^2	=	81		8 + 1 = 9
45		45^2	=	2,025		20 + 25 = 45
55		55^2	=	3,025		30 + 25 = 55
99		99^2	=	9,801		98 + 01 = 99
297		297^2	=	88,209		88 + 209 = 297
703		703^2	=	494,209		494 + 209 =703
999		999^2	=	998,001		998 + 001 = 999
2,223		$2{,}223^2$	=	4,941,729		494 + 1,729 = 2,223
2,728		2728^2	=	7,441,984		744 + 1,984 = 2,728
4,879		$4{,}879^2$	=	23,804,641		238 + 04,641 = 4,879
4,950		$4{,}950^2$	=	24,502,500		2,450 + 2,500 = 4,950
5,050		$5{,}050^2$	=	25,502,500		2,550 + 2,500 = 5,050
5,292		$5{,}292^2$	=	28,005,264		28 + 005,264 = 5,292
7,272		$7{,}272^2$	=	52,881,984		5,288 + 1,984 = 7,272
7,777		$7{,}777^2$	=	60,481,729		6,048 + 1,729 = 7,777
9,999		999^2	=	99,980,001		9,998 + 0,001 = 9,999
17,344		$17{,}344^2$	=	300,814,336		3,008 + 14,336 = 17,344
22,222		$22{,}222^2$	=	493,817,284		4,938 + 17,284 = 22,222

Figure 1-22

Other Kaprekar numbers are: 38,962; 77,778; 82656; 95,121; 99,999; 142,857; ...; 538,461; 857,143; ....

We also have such numbers as Kaprekar triples, where the cube of the number is partitioned into three numbers whose sum is the original number. For example, the number 45 meets the criteria of a Kaprekar triple, since $45^3 = 91{,}125$, and $9 + 11 + 25 = 45$. Some other Kaprekar triples are 1, 8, 10, 297, and 2322.

Earlier, we showed that 297 is a Kaprekar number, and now we show how it is also a Kaprekar triple: $297^3 = 26{,}198{,}073$, and $26 + 198 + 073 = 297$.

The smallest 10-digit Kaprekar number is 1,111,111,111, where the square of this number is the palindromic number 1,234,567,900,987,654,321, and where the sum of its parts is also a palindromic number 123,456,790 + 0,987,654,321 = 1,111,111,111, thus making it a Kaprekar number. Furthermore, the number 22,222,222,222,222 is another Kaprekar number since $22{,}222{,}222{,}222{,}222^2 = 493{,}827{,}160{,}493{,}817{,}283{,}950{,}617{,}284$. Then $4{,}938{,}271{,}604{,}938 + 17{,}283{,}950{,}617{,}284 = 22{,}222{,}222{,}222{,}222$.

Amicable Numbers

Amicable numbers are two numbers where the sum of the proper divisors (divisors excluding the original number) of one number equals the second number, *and* the sum of the proper divisors of the second number equals the first number. The smallest pair of amicable numbers is 220 and 284.

The divisors of **220** (other than 220 itself) are 1, 2, 4, 5, 10, 11, 20, 22, 44, 55, and 110.

Their sum is $1 + 2 + 4 + 5 + 10 + 11 + 20 + 22 + 44 + 55 + 110 =$ **284**. The divisors of **284** (other than 284 itself) are 1, 2, 4, 71, and 142, and their sum is $1 + 2 + 4 + 71 + 142 =$ **220**. This shows that 220 and 284 are amicable numbers.

Another pair of amicable numbers is 17,296 and 18,416. Its discovery is usually attributed to the French mathematician, Pierre de Fermat (1607–1665), although there is evidence that this discovery was anticipated by the Moroccan mathematician Ibn al-Banna al-Marrakushi al-Azdi (1256–c.1321). The sum of the proper divisors of 17,296 is

$$1 + 2 + 4 + 8 + 16 + 23 + 46 + 47 + 92 + 94 + 184 + 188 + 368 + 376 + 752 + 1{,}081 + 2{,}162 + 4{,}324 + 8{,}648 = 18{,}416.$$

The sum of the proper divisors of 18,416 is $1 + 2 + 4 + 8 + 16 + 1{,}151 + 2{,}302 + 4{,}604 + 9{,}208 = 17{,}296$. Thus, they, too, are truly amicable numbers.

The French mathematician René Descartes (1596–1650) found another pair of amicable numbers: 9,363,584, and 9,437,056. By 1747, the Swiss mathematician, Leonhard Euler (1707–1783) discovered 60 pairs of amicable numbers, yet he seemed to have overlooked the second-smallest pair, namely, 1,184 and 1,210, which was discovered in 1866 by the 16-year-old B. Nicolò I. Paganini (1850–?). The sum of the factors of 1,184 is 1 + 2 + 4 + 8 + 16 + 32 + 37 + 74 + 148 + 296 + 592 = 1,210. And the sum of the factors of 1,210 is 1 + 2 + 5 + 10 + 11 + 22 + 55 + 110 + 121 + 242 + 605 = 1,184.

To date, over 363,000 pairs of amicable numbers have identified, yet we do not know if there are an infinite number of such pairs. The table in Figure 1-23 provides a list of the first 108 pairs of amicable numbers. Going beyond this list, we will eventually stumble on an even larger pair of amicable numbers: 111,448,537,712 and 118,853,793,424.

Amicable Numbers

	First Number	Second Number	Year of Discovery
1	220	284	ca. 500 BCE
2	1184	1210	1860
3	2620	2924	1747
4	5020	5564	1747
5	6232	6368	1747
6	10744	10856	1747
7	12285	14595	1939
8	17296	18416	ca. 1310/1636
9	63020	76084	1747
10	66928	66992	1747
11	67095	71145	1747
12	69615	87633	1747
13	79750	88730	1964
14	100485	124155	1747
15	122265	139815	1747
16	122368	123152	1941/1942
17	141664	153176	1747
18	142310	168730	1747
19	171856	176336	1747
20	176272	180848	1747
21	185368	203432	1966
22	196724	202444	1747
23	280540	365084	1966
24	308620	389924	1747
25	319550	430402	1966
26	356408	399592	1921
27	437456	455344	1747
28	469028	486178	1966

(*Continued*)

(*Continued*)

29	503056	514736	1747
30	522405	525915	1747
31	600392	669688	1921
32	609928	686072	1747
33	624184	691256	1921
34	635624	712216	1921
35	643336	652664	1747
36	667964	783556	1966
37	726104	796696	1921
38	802725	863835	1966
39	879712	901424	1966
40	898216	980984	1747
41	947835	1125765	1946
42	998104	1043096	1966
43	1077890	1099390	1966
44	1154450	1189150	1957
45	1156870	1292570	1946
46	1175265	1438983	1747
47	1185376	1286744	1929
48	1280565	1340235	1747
49	1328470	1483850	1966
50	1358595	1486845	1747
51	1392368	1464592	1747
52	1466150	1747930	1966
53	1468324	1749212	1967
54	1511930	1598470	1946
55	1669910	2062570	1966
56	1798875	1870245	1967
57	2082464	2090656	1747
58	2236570	2429030	1966
59	2652728	2941672	1921
60	2723792	2874064	1929
61	2728726	3077354	1966
62	2739704	2928136	1747
63	2802416	2947216	1747
64	2803580	3716164	1967
65	3276856	3721544	1747
66	3606850	3892670	1967
67	3786904	4300136	1747
68	3805264	4006736	1929
69	4238984	4314616	1967
70	4246130	4488910	1747
71	4259750	4445050	1966
72	4482765	5120595	1957
73	4532710	6135962	1957
74	4604776	5162744	1966
75	5123090	5504110	1966
76	5147032	5843048	1747
77	5232010	5799542	1967
78	5357625	5684679	1966
79	5385310	5812130	1967
80	5459176	5495264	1967

(*Continued*)

81	5726072	6369928	1921
82	5730615	6088905	1966
83	5864660	7489324	1967
84	6329416	6371384	1966
85	6377175	6680025	1966
86	6955216	7418864	1946
87	6993610	7158710	1957
88	7275532	7471508	1967
89	7288930	8221598	1966
90	7489112	7674088	1966
91	7577350	8493050	1966
92	7677248	7684672	1884
93	7800544	7916696	1929
94	7850512	8052488	1966
95	8262136	8369864	1966
96	8619765	9627915	1957
97	8666860	10638356	1966
98	8754130	10893230	1946
99	8826070	10043690	1967
100	9071685	9498555	1946
101	9199496	9592504	1929
102	9206925	10791795	1967
103	9339704	9892936	1966
104	9363584	9437056	ca. 1600/1638
105	9478910	11049730	1967
106	9491625	10950615	1967
107	9660950	10025290	1966
108	9773505	11791935	1967

Figure 1-23

Readers who wish to search for additional amicable numbers might want to use the following method. Consider the numbers $a = 3 \times 2^n - 1$, $b = 3 \times 2^{n-1} - 1$, and $c = 3^2 \times 2^{2n-1} - 1$, where n is an integer ≥ 2. If a, b, and c are prime numbers, then $2^n \times a \times b$ and $2^n \times c$ are amicable numbers. For $n \leq 200$, only $n = 2$, 4, and 7 generate prime numbers for a, b and c.

Inspecting the list of amicable numbers in Figure 1-23, we notice that each pair is either a pair of odd numbers or a pair of even numbers. To date, we do not know if there is a pair of amicable numbers where one is odd and one is even. We also do not know if any pair of amicable numbers is relatively prime (that is, the numbers have no common factor other than 1). Perhaps these open questions contribute to our continued fascination with amicable numbers.

Happy and Unhappy Numbers

It is not clear how the term *happy number* evolved. A popular origin story tells that the British mathematician Reg Allenby's daughter came home from school after recognizing this unusual relationship and called them happy numbers. A number is considered a *happy number* if we can take the sum of the squares of its digits to get a second number, then take the sum of the squares of the digits of that second number to get a third number, then take the sum of the squares of the digits of the third number, and continue repeating this process until we get to the number 1. When this scheme is applied to a number and does not eventually generate 1, that number is called an *unhappy number*. Unhappy numbers processed in this will eventually end up in a loop and continue through the cycle.

Let's consider one of the happy numbers, say 13, and follow this process of taking the sum of the squares of the digits continuously.

$$1^2+3^2=10$$
$$1^2+0^2=1.$$

Let's consider a slightly longer path to the number 1, by using 19.

$$1^2+9^2=82$$
$$8^2+2^2=68$$
$$6^2+8^2=100$$
$$1^2+0^2+0^2=1.$$

Figure 1-24 provides a list of the happy numbers from 1 to 1000. You might want to try to test few of these for their "happiness."

1, 7, 10, 13, 19, 23, 28, 31, 32, 44, 49, 68, 70, 79, 82, 86, 91, 94, 97, 100, 103, 109, 129, 130, 133, 139, 167, 176, 188, 190, 192, 193, 203, 208, 219, 226, 230, 236, 239, 262, 263, 280, 291, 293, 301, 302, 310, 313, 319, 320, 326, 329, 331, 338, 356, 362, 365, 367, 368, 376, 379, 383, 386, 391, 392, 397, 404, 409, 440, 446, 464, 469, 478, 487, 490, 496, 536, 556, 563, 565, 566, 608, 617, 622, 623, 632, 635, 637, 638, 644, 649, 653, 655, 656, 665, 671, 673, 680, 683, 694, 700, 709, 716, 736, 739, 748, 761, 763, 784, 790, 793, 802, 806, 818, 820, 833, 836, 847, 860, 863, 874, 881, 888, 899, 901, 904, 907, 910, 912, 913, 921, 923, 931, 932, 937, 940, 946, 964, 970, 973, 989, 998, 1000.

Figure 1-24

An example of an unhappy number is 25. All unhappy numbers will eventually reach the number 4. When that happens, this procedure will lead us into a loop and we will never reach the number 1.

$$2^2+5^2=29$$
$$2^2+9^2=85$$
$$8^2+5^2=89$$
$$8^2+9^2=145$$
$$1^2+4^2+5^2=42$$
$$4^2+2^2=20$$
$$2^2+0^2=4$$
$$4^2=16$$
$$1^2+6^2=37$$
$$3^2+7^2=58$$
$$5^2+8^2=89.$$

Thus begins another loop—repeating everything again with 89, which was reached in step three of this process.

Among the happy numbers, there are those that are also prime numbers. They are known as *happy prime numbers*. The unhappy prime numbers less than 500 are the following: 7, 13, 19, 23, 31, 79, 97, 103, 109, 139, 167, 193, 239, 263, 293, 313, 331, 367, 379, 383, 397, 409, 487.

The smallest happy number that has no zeros and yet has all the other digits is 1,234,456,789. Then there is the smallest happy number that has no zeros and is also palindromic: 13,456,789,298,765,431.

Chapter 2

Arithmetic Marvels

Before the advent of electronic calculators, one had to learn and practice arithmetic algorithms so that they could be done efficiently and without much thought. Unfortunately, this activity was not taught with the thought of motivating students to like the subject, but to know it and know it well. There are lots of arithmetic shortcuts that most people do not know. When they learn them, their reaction is pure wonderment. The aim of this chapter is to introduce the reader to these aspects of arithmetic that have unfortunately been neglected in over 90% of schooling. Aside from the gee-whiz reaction, the techniques presented should change the perception of arithmetic as a mere academic exercise. The reader will be exposed to the wonderment that mathematics holds and that has been kept secret for too long.

We begin with simple aspects of common numbers. The four basic algorithms in arithmetic—addition, subtraction, multiplication, and division—will be addressed from unusual circumstances. This will give the reader insight not only into more clever ways of calculating but also into obscure secrets that highlight the power of mathematics. After we discuss various alternatives to appreciate arithmetic functions, we will focus on determining whether a number is divisible by another number without performing the division. We will present clever procedures to vastly simplify working with percents and averages. We will also alert the reader about common mistakes often made in arithmetic, some of which will be eye-opening and prove to be useful techniques. At the conclusion, we will also show some entertaining aspects of addition, known as alphametics. The intent of this chapter is not only to make the reader aware of aspects of arithmetic that have been largely ignored by the general public, but also to provide tools for enhancing arithmetic calculations and appreciating the power and beauty of mathematics.

Clever Addition

One of the most popularly repeated stories from the history of mathematics is the tale of the famous mathematician Carl Friedrich Gauss (1777–1855). When he was ten years old, he mentally added the numbers from 1 to 100 in response to a busy-work assignment given by the teacher. Although it is a cute story and generally gets a very favorable reaction, it happens to provide for us a neat little technique for adding numbers in an arithmetic sequence.

An arithmetic sequence is a list of numbers that has a common difference between consecutive numbers. For example, 2, 4, 6, 8, 10, ... is an arithmetic sequence, as are each of the following:

$$5, 10, 15, 20, 25, 30, \ldots$$
$$7, 11, 15, 19, 23, 27, \ldots$$

Perhaps the simplest arithmetic sequence is: 1, 2, 3, 4, 5, ... These numbers are called the natural numbers.

When Gauss' teacher needed some time to do clerical work, he tasked his students to find the sum of the numbers from 1 to 100. As the class sequentially added the numbers $1+2+3+4+5 \cdots +98+99+100$, giving the teacher time to do his work at his desk, Gauss got the answer in less than a minute. The story goes that the teacher ignored Gauss' response and told him to keep working. When the rest of the class completed the assignment, Gauss was the only one to get the right answer. What Gauss did to get the sum of the first 100 natural numbers, without writing a single number, was not to add the numbers in the order in which they appear, but rather to add them in the following way: the first plus the last, the second plus the next-to-last, the third plus the third from last, and so on. Without writing anything, Gauss realized that

$$\begin{aligned} 1+100&=101\\ 2+99&=101\\ 3+98&=101\\ 4+97&=101\\ &\vdots\\ 49+52&=101\\ 50+51&=101. \end{aligned}$$

Notice each pair has the same sum of 101, and so the sum of these fifty pairs of numbers is $50 \times 101 = 5{,}050$.

From this example we can derive a useful formula for adding numbers in an arithmetic sequence. In other words, we added the first and the last and multiplied this sum by one-half the number of members of the sequence. We can generalize this and get a formula for an arithmetic series of n terms, where a is the first term and l is the last term, so that the $\text{Sum} = \frac{n}{2}(a+l)$.

Here is an example of how simple it is to derive a very useful mathematical formula, one based on a lovely pattern that wasn't obvious from the start.

Checking Addition

There is a clever way to check addition without simply repeating the process and hoping to get the same result. The process is sometimes referred to as "casting out nines." Suppose you would like to check the following addition:

$$\begin{array}{r} 1234\\ 5365\\ 2384\\ 7103\\ \underline{9876}\\ 25962 \end{array}$$

We take the sum of the digits of each of the numbers and reduce these sums by multiples of 9. The same procedure is done with the sum. You may notice the sum of the sums also plays a role here in easing the procedure.

$$\begin{array}{r}1234 \to 10 \to 1\\5365 \to 19 \to 1\\2384 \to 17 \to 8\\7103 \to 11 \to 2\\\underline{9876} \to 30 \to 3\\25962 \to 24 \to [6].\end{array}$$

We then find the sum of the results from the addends $1 + 1 + 8 + 2 + 3 = 15$, from which we take out 9 to get 6. Since this is the same as the 6 that we obtained from the digit sum of 25,962 minus a multiple of 9 (namely, 18), we know that the addition is correct.

To secure your understanding, consider another example with larger numbers. We need to check if this addition is correct:

$$\begin{array}{rl}23567 \to 23 \to 5 & \\51097 \to 22 \to 4 & \\35891 \to 26 \to 8 & \to\to\to 5+4+8+5+8=30 \to [3]\\28643 \to 23 \to 5 & \\\underline{85301} \to 17 \to 8 & \\224499 \to 30 \to [3]. & \end{array}$$

We notice that the sum of the digits can further be summed to get the sequence $5 + 4 + 8 + 5 + 8 = 30$, and removing three 9s leaves us with 3. This is precisely the result of the sum of the digits of 224,499 minus a multiple of 9. Therefore, the addition is correct.

A Clever Way to Multiply by 11

Here is a technique that will truly give you a powerful device. With this technique, multiplying by 11 can be significantly simpler than using a calculator, and yet, this technique is hardly ever presented in school. What a shame! To multiply by 11 as fast as you can, write the number you are multiplying by 11. For example, suppose you would like to multiply 23×11. All you need to do is add the digits, $2 + 3 = 5$, and place that digit 5 between the two digits in the number 23 to get 253. What could be easier! We can state this method as follows:

> *To multiply a two-digit number by 11, just add the two digits and place this sum between the two digits. If the sum is greater than 9, add the 1 to the previous tens digit.*

Just for practice, let's try using this technique to multiply 54 by 11. According to the new rule, add 5 and 4 and place the sum 9 between the 5 and 4 to get 594. This trick can get a bit more difficult when the sum of the two digits is a two-digit number. We no longer have a single digit to place between the two original digits. So, if the sum of the two digits is greater than 9, we place the units digit of that sum between the two digits of the number being multiplied by 11 and "carry" the tens digit to the hundreds digit of the multiplicand. Let's try it with 78×11. We find that the sum of the two digits is $7 + 8 = 15$. Therefore, we place the 5 between the 7 and 8, add the 1 to the 7, and get [7 + 1][5][8], or 858.

You may legitimately ask if the rule holds when 11 is multiplied by a number with more than two digits. We consider now a larger number, such as 12,345, and extend our method for multiplying by 11. Here, we begin at the right-side digit and add every pair of digits going from right to left.

$$1[1 + 2][2 + 3][3 + 4][4 + 5]5 = 135{,}795.$$

Suppose we now combine the skills we have garnered through this trick of multiplying by 11 and apply it to a number. Remember, if the sum of two digits is greater than 9, use the procedure described earlier: place the units digit of this two-digit sum appropriately and carry the tens digit to the next place. To become an expert at this procedure, we will now perform a complicated variation. Let us multiply

56,789 by 11. This may be a little bit tedious, and perhaps somewhat less realistic for common use, but we will show it here merely to demonstrate the extension of this multiplication technique. Follow along as we do this step-by-step in Figure 2-1.

5[5+6][6+7][7+8][8+9]9	Add each pair of digits between the end digits
5[5+6][6+7][7+8][17]9	Add 8 + 9 = 17
5[5+6][6+7][7+8+*1*][7]9	Carry the 1 (from the 17) to the next sum
5[5+6][6+7][16][7]9	Add 7 + 8 + 1 = 16
5[5+6][6+7+*1*][6][7]9	Carry the 1 (from the 16) to the next sum
5[5+6][14][6][7]9	Add 6 + 7 + 1 = 14
5[5+6+*1*][4][6][7]9	Carry the 1 (from the 14) to the next sum
5[12][4][6][7]9	Add 5 + 6 + 1 =12
5+1[2][4][6][7]9	Add 1 + 5 = 6 to get the answer: 624,679

Figure 2-1

This method for multiplying by 11 could be a highlight in your repertoire of arithmetic skills. Not only will others be impressed with the cleverness of the method, but they may also appreciate knowing this shortcut.

Multiplication by Divisors of Powers of 10

Multiplying by factors (or divisors) of powers of 10 is a bit more involved than merely multiplying by powers of 10, but, in many cases, it can be done mentally. Let's begin by multiplying 16 by 25 (a factor of 100). Since $25 = \frac{100}{4}$, we can write this as

$$16 \times 25 = 16 \times \frac{100}{4} = \frac{16}{4} \times 100 = 4 \times 100 = 400.$$

Rather simple! All we did was divide the multiplicand by four and then multiply by 100, which means appending two zeros to the number. Further practice can be helpful, so here are some more examples:

38×25 can be written as

$$38 \times \frac{100}{4} = \frac{38}{4} \times 100 = \frac{19}{2} \times 100 = 9.5 \times 100 = 950.$$

1.7×25 can be written as

$$\frac{17}{10} \times \frac{100}{4} = \frac{17}{4} \times \frac{100}{10} = 4.25 \times 10 = 42.5.$$

(In the previous example, we used the conversion $\frac{17}{4} = \frac{16}{4} + \frac{1}{4} = 4 + 0.25 = 4.25$.)

In some cases, these tricks are more entertaining than useful, but do give some interesting insights into why we learned various arithmetic procedures in school.

In an analogous fashion, we can multiply by 125, since $125 = \frac{1000}{8}$. Here are some examples of multiplication by 125—mentally!

$$32 \times 125 = 32 \times \frac{1000}{8} = \frac{32}{8} \times 1000 = 4 \times 1000 = 4000$$

$$78 \times 125 = 78 \times \frac{1000}{8} = \frac{78}{8} \times 1000 = \frac{39}{4} \times 1000 = 9.75 \times 1000 = 9750$$

$$3.4 \times 125 = 3.4 \times \frac{1000}{8} = \frac{3.4}{8} \times 1000 = 0.425 \times 1000 = 425.$$

When multiplying by 50, you can use $50 = \frac{100}{2}$, or when multiplying by 20, you can use $20 = \frac{100}{5}$. Practice with these special numbers will help the audience (you) perform many calculations faster than the time it takes to find and turn on your calculator!

A Nifty Method to Multiply Two Numbers

Let's delve right into this unusual method of multiplying two-digit numbers with some examples and then see why this technique works. Consider the multiplication 95×97. The following steps can be done mentally (naturally, with some practice):

Step 1:	$95 + 97$	$= 192$
Step 2:	Delete the hundreds digit	$= 92$
Step 3:	Tag two zeros onto the number	$= 9200$
Step 4:	$(100 - 95) \times (100 - 97) = 5 \times 3$	$= 15$
Step 5:	Add the last two numbers	$= 9215$, which is the product being sought!

Here is another example of using this technique to multiply two two-digit numbers:

$93 \times 96 = \ldots$

$93 + 96 = 189$

$\not{1}89$ (Delete the hundreds digit.)

Tag on two zeros $= 8900$.

Then add: $(100 - 93) \times (100 - 96) = 7 \times 4 = 28$ to the previously obtained number to get 8928, which is what we sought as $8928 = 93 \times 96$.

This technique also works when seeking the product of two numbers that are further apart, such as:

89×73
$89 + 73 = 162$
~~1~~62 (Delete the hundreds digit.)
Tag on two zeros = 6200.

Then add: $(100 - 89) \times (100 - 73) = 11 \times 27 = 297$ to get 6497, the product of 89×73.

For those who might be curious as to why this unusual technique works, we can use simple algebra that will justify it. We begin with the two two-digit numbers $(100 - a)$ and $(100 - b)$ (where $0 < a$, and $b < 100$).

Step 1: $(100 - a) + (100 - b) = 200 - a - b$

Step 2: Delete the hundreds digit—which means subtracting 100 from the number:

(1) $(200 - a - b) - 100 = 100 - a - b$

Step 3: Tag on two zeros, which means multiply by 100:

$(100 - a - b) \times 100 = 10{,}000 - 100a - 100b$

Step 4: $a \times b$

Step 5: Add the last two results from steps 3 and 4, and then do some factoring:

$$
\begin{aligned}
&10{,}000 - 100a - 100b + ab \\
&= 100(100 - a) - (100b - ab) \\
&= 100(100 - a) - b(100 - a) \\
&= (100 - a)(100 - b), \text{ which is what we set out to show.}
\end{aligned}
$$

Now you just need to practice this trick in order to master it!

A Quick Multiplication

There are people who are very adept at rapid multiplication. Here is one of the procedures used to quickly and impressively multiply a pair of two-digit numbers. It is done as follows:

- Begin by multiplying the units digits of the two numbers. If a two-digit number results, then write the units digit and carry the tens-digit to the next step.
- Next, multiply the units digit of one number by the tens digit of the other number, and then multiply the tens digit of the first number by units digit of the second number and then add the two products. Then add any number carried over from the previous step. Place the units digit of the result to the left of the previously obtained single-digit and carry the tens digit to be added to the next step's calculation.
- Lastly, multiply the two tens-digits and add the carried over number from the previous step. Place this result to the left of the two previously obtained numbers, and you will have your final result of the multiplication.

At first sight this appears to be a rather complicated procedure, but after we do an example, the procedure will become simple, clear, and quick. As our example, let us use this procedure to multiply 59×38.

- First, we will multiply the units digits of the two numbers to be multiplied: $9 \times 8 = 72$. This will have given us the units digit of our ultimate answer, namely the **2**, and we will carry the 7 to the next step.
- We will now multiply the units and tens digits of the two numbers: $5 \times 8 = 40$, and $9 \times 3 = 27$. We now add these two numbers $40 + 27 = 67$ and then add the 7 carried from the previous step to get $67 + 7 = 74$. We place the **4** to the left of the previously obtained **2** and carry the 7 to the next step. So far, we have the tens and units digits of our sought-after product, namely, **42**.
- We now multiply the two tens-digits to get $3 \times 5 = 15$, and add the carried 7 to get $15 + 7 = 22$, which we now place to the left of the two previously obtained final digits to get 2242, which is the product of 59×38.

A quick review of this procedure is simply to multiply the pair of units digits, and the cross product of tens and units digits, and then the pair of tens digits—each time carrying the tens digit as appropriate. To be a successful user of this technique you will need to practice this procedure with a number of examples so that your calculating speed will be impressive. An ambitious reader may want to extend this multiplication technique to multiplying two-digit numbers by three-digit numbers, and three-digit numbers by other three-digit numbers.

The Russian Peasant's Method of Multiplication

It is said that Russian peasants used a rather strange, perhaps even primitive, method to multiply two numbers. It is actually quite simple, if somewhat cumbersome. Let's take a look at it.

Consider the problem of finding the product of 43×92. We begin by setting up a chart of two columns with the two factors in the first row, as shown in Figure 2-2, where 43 and 92 head up the columns. One column will be formed by doubling each number to get the next number, while the other column will take half the number and drop the remainder. For convenience, our first column will be the doubling column, the second column will be the halving column. Notice that when we divide in half the odd number 23 (the third number in the second column) we get 11 with a remainder of 1, and we simply drop the 1. The rest of this halving process should now be clear.

Doubles	*Halves*
43	92
86	46
172	**23**
344	**11**
688	**5**
1376	2
2752	**1**

Figure 2-2

Find the odd numbers in the halving column (the right column), then get the sum of the partner numbers in the doubling column (the left column). These are highlighted in bold type. This sum gives you the originally required product of 43 and 92. In other words with the Russian peasant's method we get $43 \times 92 = 172 + 344 + 688 + 2752 = 3956$.

In the example above, we chose to make the first column the doubling column and the second column the halving column. We could also have done this Russian peasant's method by halving the numbers in the first column and doubling those in the second, as shown in Figure 2-3.

Halves	*Doubles*
43	**92**
21	**184**
10	368
5	**736**
2	1472
1	**2944**

Figure 2-3

To complete the multiplication, we find the odd numbers in the halving column (in bold type) and then get the sum of their partner numbers in the second column (the doubling column). This gives us $43 \times 92 = 92 + 184 + 736 + 2944 = 3956$.

You are not expected to do your multiplication in this high-tech era by copying the Russian peasant's method. However, it's fun to observe how this primitive system of arithmetic actually does work. Explorations of this kind are not only instructive but entertaining. Here you see what was done in the above multiplication algorithm.

$$
\begin{aligned}
{}^*43 \times 92 &= (21 \times 2 + 1)(92) &&= 21 \times 184 &&+ \quad 92 = 3956 \\
{}^*21 \times 184 &= (10 \times 2 + 1)(184) &&= 10 \times 368 &&+ 184 = 3864 \\
10 \times 368 &= (5 \times 2 + 0)(368) &&= 5 \times 736 &&+ \quad 0 = 3680 \\
{}^*5 \times 736 &= (2 \times 2 + 1)(736) &&= 2 \times 1472 &&+ 736 = 3680 \\
2 \times 1472 &= (1 \times 2 + 0)(1472) &&= 1 \times 2944 &&+ \quad 0 = 2944 \\
{}^*1 \times 2944 &= (0 \times 2 + 1)(2944) &&= \quad 0 &&+ \underline{2944} = 2944. \\
& && && \quad 3956
\end{aligned}
$$

For those familiar with the binary system (base 2), one can also explain this Russian peasant method with the following representation:

$$\begin{aligned}(43)(92) &= (1\times2^5+0\times2^4+1\times2^3+0\times2^2+1\times2^1+1\times2^0)(92)\\ &= 2^0\times92+2^1\times92+2^3\times92+2^5\times92\\ &= 92+184+736+2944\\ &= 3956.\end{aligned}$$

Whether or not you fully understand the Russian peasant's method of multiplication, you should at least now have a deeper appreciation for the multiplication algorithm you learned in school, even though most people today multiply with a calculator. There are many other multiplication algorithms, yet the one shown here is perhaps one of the strangest. It is through this strangeness that we can appreciate the powerful consistency of mathematics that allows us to conjure up such an algorithm.

Speed Multiplying by 21, 31, 41, and Others

Much of our arithmetic depends on our use of algorithms. When we learned these algorithms in the early grades of our schooling, we used them without much concern for why they work. In today's technologically-advanced world, we find less need for these algorithms. You might enjoy trying to determine why these algorithms actually work. By considering the following "new" algorithm, one you may not have seen before, you have the opportunity to familiarize yourself with an unfamiliar algorithm and then try to see if you can explain why it works. Once you feel comfortable with the algorithm you can easily extend it to other numbers of this pattern. The examples provided show the algorithm working for the numbers 21, 31, and 41. You should have no trouble multiplying by 51, 61, etc. Going through this little exercise should give you a better appreciation for the standard algorithms that we are required to learn by rote in school.

To Multiply by 21: Double the Number, Then Multiply by 10 and Add the Original Number

For example, to multiply 37 by 21:

Double 37 yields 74, multiply by 10 to get 740, and then add the original number 37 to get 777.

To Multiply by 31: Triple the Number, Then Multiply by 10 and Add the Original Number

For example, to multiply 43 by 31:

Triple 43 yields 129, multiply by 10 to get 1290, and then add the original number 43 to get 1333.

To Multiply by 41: Quadruple the Number, Then Multiply by 10 and Add the Original Number

For example, to multiply 47 by 41:

Quadruple 47 yields 188, multiply by 10 to get 1880, and then add the original number 47 to get 1927.

By now, you should recognize the pattern and be able to extend its rule further.

This procedure will hold for other two-digit numbers, such as 51, 61, 71, 81, and 91. It even works with numbers such as 101, 111, 121, 131, etc. as long as you do the first multiplication for each of these by 10, 11, 12, and 13, respectively, and follow the procedure by multiplying the result by 10 and adding the original number.

The Product of Two Negative Numbers

There are times when a "math person" is asked to explain a common concept that is commonly accepted without much justification. The mathematics expert could well be asked to explain the reason why the product of two negative numbers is a positive number. We know that (in English) when two negative statements are made in one sentence the result is positive, such as the sentence "She does not take no money out of her bank account." In effect, this translates to mean "She does take money from her bank account." In mathematics we are often faced with the question: how can we demonstrate that the product of two negatives is a positive? If we wanted to explain this algebraically, here is a tricky way using the distributive property.

We begin with an equality to which 0 is added, leaving the equality intact:

$$(-a)(-b)=(-a)(-b)+(0)(b).$$

Then substituting $0=(a-a)$:

$$(-a)(-b)=(-a)(-b)+(-a+a)(b).$$

Applying the distributive property give us:

$$(-a)(-b)=(-a)(-b)+(-a)(b)+(a)(b).$$

Once again, the distributive property yields:

$$(-a)(-b)=(-a)(-b+b)+(a)(b).$$

Then it follows that:

$$(-a)(-b)=(-a)(0)+(a)(b).$$

And we reach our desired conclusion:

$$(-a)(-b)=(a)(b).$$

There are many other ways to show that the product of two negatives is a positive; for instance, if you do not take apples out of a basket of apples, then you will have apples in the basket. The math expert may also decide to demonstrate this property by showing a progression of multiplications such as:

$$-3 \times 2 = -6$$
$$-3 \times 1 = -3$$
$$-3 \times 0 = 0$$
$$-3 \times -1 = +3$$
$$-3 \times -2 = +6, \text{ and so on.}$$

However, the previous procedure might be considered more mathematically elegant and convincing. Here we let the reader choose which explanation suits best.

Squaring a Multiple of 5

You might want to challenge your audience to square the number 85 mentally. Initial reactions are usually "Are you kidding?" Well, here we will show you a way to perform this trick rather easily. Let's consider a general case of a two-digit number $(10a + 5)$, which is a multiple of 5. When we square this number, we get $(10a + 5)^2 = 100a^2 + 100a + 25 = a(a + 1)(100) + 25$. This tells us to take the tens digit and multiply it by the tens digit +1, multiply that by 100, and add 25. We can do this to find the value of 85^2. All we need to do, following this technique, is to multiply the tens digit, which is 8, by $8 + 1 = 9$, and then multiply that by 100 and add 25, as shown here: $85^2 = (8)(9)(100) + 25 = 7225$. With a little practice this can go quite quickly, and you will have another useful technique in your arithmetic toolbox.

Now that we have provided some useful techniques for multiplication, which will provide insights into arithmetic processes that can also shorten calculation time. We are now ready to take up the process of division. We begin by considering how to determine whether a given number is divisible by various other numbers.

A Method to Determine When a Number Is Divisible by 11

While we previously considered a clever way of multiplying by 11, we should also see if there is an insightful way to determine if a given number is divisible by 11. Thus, you will be fortified with another novelty considering the number 11, which owes its special status to being 1 greater than 10, the base of our number system. The need to determine if a number is divisible by 11 can arise at the oddest times. If you have a calculator at hand, the problem is easily solved. But that is not always the case. Besides, there is such a clever method for testing for divisibility by 11 that it is worth knowing just for its charm. The method is quite simple:

If the difference of the sums of the alternate digits is divisible by 11, then the original number is also divisible by 11.

This sounds more complicated than it is. Let us take this method a piece at a time. "The sums of the alternate digits" means that you begin at one end of the number, take the first, third, fifth, etc. digits, and add them. Then add the remaining (even placed) digits. Subtract the two sums and inspect for divisibility by 11. If the resulting number is divisible by 11, then the original number is divisible by 11. And the reverse is also true; that is, if the number reached by subtracting the two sums is *not* divisible by 11, then the original number is also *not* divisible by 11.

This is probably best demonstrated through an example. Suppose we test the number 768,614 for divisibility by 11. The sums of the alternate digits are: $7 + 8 + 1 = 16$, and $6 + 6 + 4 = 16$. The difference of these two sums is $16 - 16 = 0$, which is divisible by 11. (Remember, $\frac{0}{11} = 0$.) Therefore, we can conclude that 768,614 is divisible by 11.

Another example might be helpful to firm up your understanding of this procedure. To determine if 918,082 is divisible by 11, we need to find the sums of the alternate digits:

$$9 + 8 + 8 = 25, \text{ and } 1 + 0 + 2 = 3.$$

Their difference is 25 – 3 = 22, which is divisible by 11, and so the number 918,082 is divisible by 11. Now just practice this method so that you can demonstrate it more easily and at the same time impress your friends by stressing the power and consistency of mathematics.

In case you are asked why this method works as it does, we offer a brief justification. Consider the number *ab,cde*, whose value can be expressed as

$$\begin{aligned}N &= 10^4 a + 10^3 b + 10^2 c + 10d + e\\ &= (11-1)^4 a + (11-1)^3 b + (11-1)^2 c + (11-1)d + e\\ &= [11M + (-1)^4]a + [11M + (-1)^3]b + [11M + (-1)^2]c + [11 + (-1)]d + e,\\ &\quad \text{(where } 11M \text{ represents a multiple of } M)\\ &= 11M[a+b+c+d] + a - b + c - d + e,\end{aligned}$$

which implies that since the first part of this value of N, namely, $11M[a+b+c+d]$, is already a multiple of 11, the divisibility by 11 of N depends on the divisibility of the remaining part, $a - b + c - d + e = (a + c + e) - (b + d)$, which is the difference of the sums of the alternate digits.

At Times the Brain Can Be More Effective Than a Calculator

It is very motivational to see how mental calculation can be more effective than taking the time to input information in a calculator. For example, whether a given number is divisible by 5 or 10 can be easily determined just by looking at the number to see if the units digit is a 0 or a 5. Similarly, we know that even numbers are divisible by 2. But how do you determine whether a number is divisible by 4? There are even numbers that obviously are not divisible by 4, such as the number 354. A clever mental method for determining whether a number is divisible by 4 is to simply look at the last two digits of the number and see if that two-digit number is divisible by 4. If it is, then the original number is also divisible by 4. Consider the number 3784, where the last two digits are the number 84, which is clearly divisible by 4.

Therefore, the original number, 3784, is also divisible by 4. This can be extended for other powers of 2, such as $2^3 = 8$: we consider the last three digits of the number, and when that three-digit number is divisible by 8, then the entire number is also divisible by 8. If we look at the number 213,648, we find that the number 648 is divisible by 8 (since $648 = 8 \times 81$). Therefore, 213,648 is divisible by 8.

Determining Whether a Number Is Divisible by 3 or 9

Folks are always interested in how useful math can be. Suppose you are in a restaurant with two friends and receive a check of $71.22. You want to add a $14 tip, but you also want the complete sum to be split into three equal parts. Wouldn't it be nice if there were some mental arithmetic method for determining this? Well, here comes mathematics to the rescue. We are going to provide you with a simple technique to determine if a number is divisible by 3 and, as an extra bonus, also divisible by 9.

However, since smartphones have calculators, when asked to determine if a given number is divisible by 3 or 9, most people would be tempted to rely on a calculator, especially if the number is large. Yet, there is also a mental method to determine these divisibilities.

> *If the sum of the digits is divisible by 3 or 9, then the number is divisible by 3 or 9.*

For example, consider the number 795,324. This number has a digit sum of $7 + 9 + 5 + 3 + 2 + 4 = 30$, which is divisible by 3 but not 9. Therefore, the original number 795,324 is divisible by 3 and not 9. Whereas the number 89,784 has a digit sum of $8 + 9 + 7 + 8 + 4 = 36$, which is divisible by 3 and 9, and therefore the original number, 89,784, is divisible by 3 and 9.

As another example, is the number 27,987 divisible by 3 or 9? The sum of the digits is $2 + 7 + 9 + 8 + 7 = 33$, which is divisible by 3 but not by 9; therefore, the number 27,987 is divisible by 3 and not by 9.

Now that you are expert at determining if a number is divisible by 3 or 9, we can go back to the original question about the divisibility of the restaurant bill of \$71.22 + 14.00 = \$85.22. Can it be divided into three equal parts? Because 8 + 5 + 2 + 2 = 17, and 17 is not divisible by 3, then \$85.22 is not divisible by 3. You will have to modify the tip to get everyone to pay the same amount.

Appreciating mathematics involves not just seeing how it can be helpful but also *why* the helpful trick actually works. To explain why this technique works, consider the number *ab,cde,* whose value can be expressed as follows: (where $9M$ indicates a multiple of 9)

$$
\begin{aligned}
N &= 10^4 a + 10^3 b + 10^2 c + 10d + e \\
&= (9+1)^4 a + (9+1)^3 b + (9+1)^2 c + (9+1)d + e \\
&= [9M + (1)^4]a + [9M + (1)^3]b + [9M + (1)^2]c + [9 + (1)]d + e \\
&= 9M[a+b+c+d] + a + b + c + d + e,
\end{aligned}
$$

which implies that divisibility by 9 of the original number N depends on the divisibility of $a + b + c + d + e$, which is the sum of the digits. The same argument can be used for 3.

Divisibility by Prime Numbers

It has always been most perplexing to establish mental calculating methods for determining divisibility by prime numbers. This is especially true for testing divisibility by 7, which follows a series of nifty divisibility schemes for the numbers 2 through 5. (By the way, the method to determine divisibility by 6 is simply to apply the technique for divisibility by 2 and by 3; both must hold true for a number to be divisible by 6.)

As you will soon see, the divisibility tests for some prime numbers are almost as cumbersome as an actual division algorithm. That being said, they are fun, and, believe it or not, can be used to impress others. They often stir up curiosity about why these methods, which actually work, weren't taught in the lower grades, where the focus is on arithmetic. So, let us consider divisibility methods for other prime numbers.

Test for Divisibility by 7

We begin by considering the clever procedure to determine if a given number is divisible by 7. Then, as we inspect the procedure, we can see how it can be generalized and applied for other prime numbers. The procedure to test for divisibility by 7 is as follows:

> *Delete the last digit (units digit) from the given number and then subtract twice this deleted digit from the remaining number. If the result is divisible by 7, the original number is divisible by 7. This process may be repeated until by simple inspection the result can be determined for divisibility by 7.*

As an example, let's test the number 876,547 for divisibility by 7. Begin with 876,547 and delete its units digit, 7, and subtract its double $2 \times 7 = 14$, from the remaining number: $87{,}654 - 14 = 87{,}640$. Since we cannot yet visually inspect the resulting number for divisibility by 7, we continue the process with the resulting number 87,640: we delete its units digit, 0, and subtract its double, still 0, from the remaining number to get $8{,}764 - 0 = 8{,}764$. This did not bring us any closer to visually checking for divisibility by 7. Therefore, we continue the process with the resulting number 8,764: we delete its units digit, 4, and subtract its double, 8, from the remaining number to get $876 - 8 = 868$. Since we still cannot visually inspect the resulting number, 868, for divisibility by 7, we continue the process with the resulting number 868: delete its units digit, 8, and subtract its double, 16, from the remaining number to get $86 - 16 = 70$, which we can easily see is divisible by 7. Therefore, the original number 876,547 is also divisible by 7, since

$$\frac{876547}{7} = 125{,}221.$$

Before continuing with our discussion of divisibility of prime numbers, it would be wise to practice this method with a few randomly selected numbers and then check your results with a calculator.

Now for the beauty of mathematics! Why does this rather strange procedure actually work? To show why it works is a wonderful thing about mathematics. This will all make sense to you after you see what is happening with this procedure. Again, this is where we can enhance the beauty of mathematics.

To justify the method of determining divisibility by 7, consider the various possible terminal digits (that you are "dropping") and the corresponding subtraction that is actually being done by dropping the last digit. In Figure 2-4, you will see that dropping the terminal digit and doubling it means that the number being subtracted in each case is a multiple of 7, that is, 21, 42, 63, 84, . . . 168, and 189 are all multiples of 7. In other words, we have taken "bundles of 7" away from the original number. Therefore, if the remaining number is divisible by 7, then so is the original number, because you have separated the original number into parts, each of which is divisible by 7. Therefore, the entire number must be divisible by 7.

Terminal digit	Number subtracted from original	Terminal digit	Number subtracted from original
1	20 + 1 = 21 = 3×7	5	100 + 5 = 105 = 15×7
2	40 + 2 = 42 = 6×7	6	120 + 6 = 126 = 18×7
3	60 + 3 = 63 = 9×7	7	140 + 7 = 147 = 21×7
4	80 + 4 = 84 = 12×7	8	160 + 8 = 168 = 24×7
		9	180 + 9 = 189 = 27×7

Figure 2-4

Test for Divisibility by 13

Here is an analogous method for divisibility by 13: *Follow the rule for testing divisibility by 7, but instead of subtracting twice the deleted last digit, subtract 9 times the deleted digit each time.*

Let's apply the method to check the number 5,616 for divisibility by 13. We begin by deleting its units digit, 6, and subtract $9 \times 6 = 54$, from the remaining number: $561 - 54 = 507$. Since we still cannot visually inspect the resulting number for divisibility by 13, we continue the process with the resulting number 507, deleting its units digit and subtracting 9 times this digit ($9 \times 7 = 63$) from the remaining number: $50 - 63 = -13$, which is divisible by 13. Therefore, the original number is also divisible by 13, since

$$\frac{5616}{13} = 432.$$

Now we might want to see how we determined the "multiplier" 9 in our method. We sought the smallest multiple of 13 that ends in a 1. That was 91, where the tens digit is 9 times the units digit. Once again, consider the various possible terminal digits and the corresponding subtractions in Figure 2-5.

Terminal digit	Number subtracted from original	Terminal digit	Number subtracted from original
1	$90 + 1 = 91 = 7 \times 13$	5	$450 + 5 = 455 = 35 \times 13$
2	$180 + 2 = 182 = 14 \times 13$	6	$540 + 6 = 546 = 42 \times 13$
3	$270 + 3 = 273 = 21 \times 13$	7	$630 + 7 = 637 = 49 \times 13$
4	$360 + 4 = 364 = 28 \times 13$	8	$720 + 8 = 728 = 56 \times 13$
		9	$810 + 9 = 819 = 63 \times 13$

Figure 2-5

In each case, a multiple of 13 is being subtracted one or more times from the original number. Hence, if the remaining number is divisible by 13, then the original number is divisible by 13.

Test for Divisibility by 17

One would probably want to see if this procedure can be carried out for further prime numbers. So, we consider the method for testing divisibility by 17, which is as follows:

> *Delete the units digit and subtract 5 times the deleted digit each time from the remaining number until you reach a number small enough to visually determine its divisibility by 17.*

We justify the method for divisibility by 17 as we did the methods for 7 and 13. Each step of the process requires us to subtract a "bunch of 17s" from the original number until we reduce the number to a manageable size, by which we can make a visual inspection of divisibility by 17. This time, the multiplier is 5, since we will be deducting multiples of 17, such as 51, 102, 153, and so on, from the original number.

As an example, suppose we wish to determine whether the number 123,456 is divisible by 17. We delete the units digit, 6, multiply it by 5 to get 30, and subtract 30 from the remaining number 12,345: 12,345 − 30 = 12,315. Since we cannot determine if 12,315 is divisible by 17, we continue the process so that we now subtract 1231 – 25 = 1206. We need to continue the process so that we now subtract 120 – 30 = 90, = which is not divisible by 17, and therefore, the original number 123,456 is also not divisible by 17.

As another example, test the number 146,795 for divisibility by 17. We begin by subtracting 14,679 – 25 = 14,654. Continuing the process, we subtract 1465 – 20 = 1445. Next step, 144 – 25 = 119. Then 11 – 45 = – 34, which is –2 × 17, and therefore, 146,795 is divisible by 17.

Test for Divisibility by Other Prime Numbers

The patterns developed in the preceding three divisibility methods (for 7, 13, and 17) enable us to produce analogous methods for testing divisibility by larger primes. Figure 2-6 presents the "multipliers" of the deleted digits for various prime divisibility tests.

To test divisibility by	7	11	13	17	19	23	29	31	37	41	43	47
Multiplier	2	1	9	5	17	16	26	3	11	4	30	14

Figure 2-6

You may want to extend this chart. It's fun, and it will improve your perception of mathematics, while at the same time expanding your toolbox of arithmetic methods.

Test for Divisibility by Composite Numbers

You may also want to extend your knowledge of divisibility rules to include composite (i.e., non-prime) numbers. The following method refers to *relatively prime* factors, not just any factors. Recall that two numbers are relatively prime if they have no common factors other than 1. Consider the following method for divisibility by composite numbers:

> *A given number is divisible by a composite number if it is divisible by each of its relatively prime factors.*

Figure 2-7 illustrates this method of checking divisibility by composite numbers.

To be divisible by	6	10	12	15	18	21	24	26	28
The number must be divisible by	2 & 3	2 & 5	3 & 4	3 & 5	2 & 9	3 & 7	3 & 8	2 & 13	4 & 7

Figure 2-7

At this juncture you have not only a comprehensive list of methods for testing divisibility, but also an interesting insight into elementary number theory. This is rarely presented in school, where it is typically absent from the regular curriculum. Yet, taking the time to enrich students with this topic further enhances their enthusiasm for mathematics by exposing them to something that is not well-known by the general public. Indeed, the unusualness of this topic further enhances an appreciation of mathematics.

Additional Tests for Divisibility by 19

Method 1: The number 19 is not usually considered a special number; however, it does lend itself to some interesting aspects. Here is an interesting test for checking a number's divisibility by 19. Take the number, delete the last digit, and then add the double of this digit to the remaining number. Continue this process until you reach a number that is obviously divisible or indivisble by 19. If the end result is divisible by 19, then the original number is divisible by 19. For example, suppose we want to determine whether the number 1387 is divisible by 19. We first delete the last digit, 7, and add twice that digit to the remaining number: $138 + 14 = 152$. Repeating this process, we get $15 + 4 = 19$; therefore, the original number 1387 is divisible by 19. Here is another example where we can determine if the number 67,982 is divisible by 19. Follow along the steps: $6798 + 4 = 6802$; $680 + 4 = 684$; $68 + 8 = 76$; $7 + 12 = 19$, which verifies that the original number 67,982 is divisible by 19.

Method 2: There is another interesting test for divisibility by 19. We have to represent the original number in the form of $100p + q$. If, and only if, $p + 4q$ is divisible by 19, then the original number is divisible by 19. Checking once again the number 1387 for divisibility by 19, we would represent that number as $(13 \times 100) + 87$. We now find that $13 + (4 \times 87) = 361$, which happens to be 19^2, and so it is divisible by 19; therefore, the original number 1387 is divisible by 19.

Method 3: Another procedure for determining whether a given number is divisible by 19 is to conform to the earlier methods of checking for divisibility by prime numbers. Thus, to test for the divisibility by 19, we delete the units digit, multiply it by 17, and then subtract that number from the remainder of the original number. Continue this process until a number results that is obviously divisible by 19 or not. For example, consider the number 34,542. To test divisibility by 19 using this procedure, we delete the units digit, 2, and subtract $17 \times 2 = 34$ from the remaining number 3454 to get 3420. Following this procedure, the resulting number from the next step is 342. Once again, by deleting the units digit, multiplying it by 17 to get 34, and subtracting that from the remaining number, 34, we get 0, which is a multiple of 19, and therefore, the original number 34,542 is divisible by 19.

Method 4: There is yet another test for divisibility by 19 that is a bit more cumbersome but can still be effective in demonstrating the power of numbers (although it's not necessarily more effective than a calculator). The process goes as follows: delete the units digit from the number you are testing and multiply the remaining number by 9, after which you subtract the units digit. Continue this process until you reach a number that you can visually recognize as either 0 or a multiple of 19. For example, suppose we want to determine if the number 152 is divisible by 19. We follow the process: $(15 \times 9) - 2 = 133$, then $(13 \times 9) - 3 = 114$, then $(11 \times 9) - 4 = 95$, then $(9 \times 9) - 5 = 76$, then $(7 \times 9) - 6 = 57$. Most people will recognize that the last three results are multiples of 19, and therefore, the original number 152 is divisible by 19, Naturally, one could have stopped much sooner, once the number arrived at was obviously divisible by 19. Although this method is a bit cumbersome and not necessarily a shortcut, it does provide another view of what can be done arithmetically with our number system.

Test for Divisibility by 101

The number 101 is the largest known prime which is 1 greater than a power of 10. There is a clever test to determine if a number is divisible by 101. Although it is a bit cumbersome, it does give some insight into the number 101. Going from right to left, separate the number into pairs of digits, and then alternate between subtracting and adding the digits. An end result of either 0 or 101 would indicate that the original number is divisible by 101. For example, let's determine if the number 854,089,043,463 is divisible by 101.

Begin by separating the number from right to left into pairs of digits: 85, 40, 89, 04, 34, 63. Then subtract and add the pairs alternately: $63 - 34 + 04 - 89 + 40 - 85 = -101$. Therefore, the original number is divisible by 101.

Divisibility by 1001

Since the number 1001 is the product of the three prime numbers 7, 11, and 13, a clever algorithm can determine if a very large number is divisible by the numbers 7, 11, and 13. This is best explained in the context of a specific situation. Consider the number 23,592, 840,010,274,535. To determine if it is divisible by 7, 11, and 13, we split it into groups of three digits from right to left, and take the sum of the second, fourth, and sixth group of three digits: $274 + 840 + 23 = 1137$. We then take the sum of the remaining groups of three digits: $535 + 010 + 592 = 1137$, and subtract the two sums: $1137 - 1137 = 0$. If the resulting difference is divisible by 7, 11, and

13 (that is, by 1001) then the original number is divisible by 1001. Here, our difference is 0; therefore, the original number is divisible by 7, 11, and 13, since 0 is divisible by any number.

A Technique to Determine Divisibility by Single-Digit Numbers

This technique enables one to easily determine divisibility by numbers 2 through 9. We simply separate the digits of our original number, multiply the tens digit by 10 minus the sought-after single-digit number, and add to it the units digit. If the result is a multiple of the desired single-digit number, then the original number is divisible by that number. Perhaps it is best to show this technique through a variety of examples. Suppose we want to determine if the number 96 is divisible by 8. For 96, we do the following: $9 \times 2 = 18 + 6 = 24$. Since 24 is divisible by 8, we can conclude that 96 is also divisible by 8. Now the question is, why did we multiply the 9×2? The 2 was selected because it is the difference between 10 and 8. The 6 we added was the original units digit. To firm up this technique, let us do a few others.

This time, we would like to determine whether the number 182 is divisible by 7. To do that, we split up the number by separating the units digit from the remainder of the number. This time, we will multiply the number excluding the units digit by 3, since $10 - 7 = 3$. We now do the following: $18 \times 3 = 54 + 2 = 56$, which is divisible by 7, and therefore, we can conclude that 182 is also divisible by 7. The 2 we added came from the original units digit.

Just to further secure your understanding of this procedure, let's consider one more example and determine if the number 687 is divisible by 7. Remember, our multiplier will once again be 3 (since $10 - 7 = 3$), so that our procedure is as follows: $68 \times 3 = 204 + 7 = 211$. Since we cannot immediately determine if the number 211 is divisible by 7, we simply repeat the procedure: $21 \times 3 = 63 + 1 = 64$, which is not divisible by 7. Therefore, the number 687 is also not divisible by 7.

An ambitious reader may wish to use simple algebra to determine why this method works.

Arithmetic Curiosities

Before we embark on further arithmetic shortcuts and techniques, we should appreciate the strange symmetry that can be expressed in arithmetic terms. This is merely to exhibit relationships that are not well-known, but present certain inspiring aspects of arithmetic. A curious reader might find other such relationships.

The Symmetry Between Multiplication and Addition

Figure 2-8 exhibits a symmetry between multiplication and addition:

$$\left(1+\frac{1}{2}\right)\times 3=\left(1+\frac{1}{2}\right)+3$$
$$\left(1+\frac{1}{3}\right)\times 4=\left(1+\frac{1}{3}\right)+4$$
$$\left(1+\frac{1}{4}\right)\times 5=\left(1+\frac{1}{4}\right)+5$$
$$\left(1+\frac{1}{5}\right)\times 6=\left(1+\frac{1}{5}\right)+6$$
$$\left(1+\frac{1}{6}\right)\times 7=\left(1+\frac{1}{6}\right)+7$$
$$\left(1+\frac{1}{7}\right)\times 8=\left(1+\frac{1}{7}\right)+8$$
$$\left(1+\frac{1}{8}\right)\times 9=\left(1+\frac{1}{8}\right)+9$$
$$\left(1+\frac{1}{9}\right)\times 10=\left(1+\frac{1}{9}\right)+10$$

Figure 2-8

Algebraically, this can be expressed as

$$\left(1+\frac{1}{n}\right)\times(n+1)=\left(1+\frac{1}{n}\right)+(n+1).$$

To justify this quality, we simply perform elementary algebra calculations:

$$\left(\frac{n+1}{n}\right)\times(n+1)=?=\left(\frac{n+1}{n}\right)+(n+1)$$

$$\frac{(n+1)(n+1)}{n}=?=\left(\frac{n+1}{n}\right)+\left(\frac{n(n+1)}{n}\right)$$

$$\frac{n^2+2n+1}{n}=?=\frac{n+1+n^2+n}{n}$$

$$\frac{n^2+2n+1}{n}=\frac{n^2+2n+1}{n}.$$

Relating Multiplication and Subtraction

There are symmetrical curiosities involving multiplication and subtraction as partners, as shown in Figure 2-9.

$$1\times\frac{1}{2}=\frac{1}{2}=1-\frac{1}{2}$$

$$2\times\frac{2}{3}=1\frac{1}{3}=2-\frac{2}{3}$$

$$3\times\frac{3}{4}=2\frac{1}{4}=3-\frac{3}{4}$$

$$4\times\frac{4}{5}=3\frac{1}{5}=4-\frac{4}{5}$$

$$5\times\frac{5}{6}=4\frac{1}{6}=5-\frac{5}{6}$$

$$6\times\frac{6}{7}=5\frac{1}{7}=6-\frac{6}{7}$$

Figure 2-9

This can be justified and explained algebraically by showing that the left side of the equation is equal to the right side.

$$n\times\frac{n}{n+1}=?=n-\frac{n}{n+1}$$

The left side: $n\times\frac{n}{n+1}=\frac{n^2}{n+1}$

The right side: $n-\frac{n}{n+1}=\frac{n(n+1)-n}{n+1}=\frac{n^2}{n+1}$.

Further Arithmetic Curiosities Comparing Division to Addition

Figure 2-10 exhibits yet another symmetry, this time between division and addition.

$$1\frac{1}{3}\div\frac{2}{3}=1\frac{1}{3}+\frac{2}{3}=2$$
$$2\frac{1}{4}\div\frac{3}{4}=2\frac{1}{4}+\frac{3}{4}=3$$
$$3\frac{1}{5}\div\frac{4}{5}=3\frac{1}{5}+\frac{4}{5}=4$$
$$4\frac{1}{6}\div\frac{5}{6}=4\frac{1}{6}+\frac{5}{6}=5$$
$$5\frac{1}{7}\div\frac{6}{7}=5\frac{1}{7}+\frac{6}{7}=6$$

Figure 2-10

Represented algebraically, we have the following:

$$\left(n+\frac{1}{n+2}\right)\div\frac{n+1}{n+2}=?=\left(n+\frac{1}{n+2}\right)+\frac{n+1}{n+2}$$

Left side: $\left(n+\frac{1}{n+2}\right)\div\frac{n+1}{n+2}=\left(n+\frac{1}{n+2}\right)\times\frac{n+2}{n+1}$

$$=\frac{n(n+2)+1}{n+2}\times\frac{n+2}{n+1}=\frac{n^2+2n+1}{n+1}=n+1$$

$$\text{Right side: } \left(n+\frac{1}{n+2}\right)+\frac{n+1}{n+2}=\frac{n(n+2)+1}{n+2}+\frac{n+1}{n+2}$$

$$=\frac{n^2+2n+1+n+1}{n+2}=\frac{n^2+3n+2}{n+2}=\frac{(n+1)(n+2)}{n+2}=n+1.$$

Next, Curious Arithmetic Relating Division and Subtraction

This time, we show a symmetry between division and subtraction in Figure 2-11.

$$4\frac{1}{2}\div 3=4\frac{1}{2}-3=1\frac{1}{2}$$
$$5\frac{1}{3}\div 4=5\frac{1}{3}-4=1\frac{1}{3}$$
$$6\frac{1}{4}\div 5=6\frac{1}{4}-5=1\frac{1}{4}$$
$$7\frac{1}{5}\div 6=7\frac{1}{5}-6=1\frac{1}{5}$$
$$8\frac{1}{6}\div 7=8\frac{1}{6}-7=1\frac{1}{6}$$

Figure 2-11

Algebraically, this appears as follows:

$$\left(n+\frac{1}{n-2}\right)\div(n-1)=?=\left(n+\frac{1}{n-2}\right)-(n-1)$$

$$\text{Left side: } \left(n+\frac{1}{n-2}\right)\div(n-1)=\frac{n(n-2)+1}{n-2}\times\frac{1}{n-1}$$

$$=\frac{n^2-2n+1}{(n-2)(n-1)}=\frac{(n-1)(n-1)}{(n-2)(n-1)}=\frac{(n-1)}{(n-2)}$$

$$\text{Right side: } \left(n+\frac{1}{n-2}\right)-(n-1)=\frac{n(n-2)+1}{n-2}-\frac{(n-1)(n-2)}{(n-2)}$$

$$=\frac{(n^2-2n+1)-(n^2-3n+2)}{n-2}=\frac{n-1}{n+2}.$$

Now that we have admired some of the strange ways that arithmetic can be perceived, we shall now investigate the various ways of doing arithmetic in a more expedient way. We will begin by admiring shortcuts for multiplication, and then the inverse, divisibility.

Maximizing a Product

While we are concerned about arithmetic nuances, it would be interesting to see how you can find the largest product formed by two five-digit numbers, where no digit is used more than once across the two numbers. Logical thinking tells us that the largest numerals must be at the left end of each number. Consider the product $86{,}420 \times 97{,}531 = 8{,}428{,}629{,}020$. Notice that the two numbers in the product use all 10 digits exactly once. You should also recognize a pattern as to how these two numbers were developed to create the largest product. Each number has descending alternating digits beginning with the largest single-digit—either 9 or 8.

Successive Percentages

Most folks consider percentage problems to be their nemesis. Such problems get particularly unpleasant when multiple percents need to be processed in the same problem. This insightful technique for solving successive percentage problems turns a one-time nemesis into a delightfully simple arithmetic algorithm with lots of useful applications. This not-very-well-known procedure should enchant you. Let's begin by considering the following problem:

> Wanting to buy a coat, Barbara is faced with a dilemma.Two competing stores next to each other carry the same brand of coat with the same list price, but with two different discount offers. Store A offers a 10% discount year-round on all its goods, but on this particular day offers an additional 20% on top of their already discounted price. Store B simply offers a discount of 30% on that day in order to stay competitive. What is the difference between the two discount options open to Barbara?

At first glance, you may assume there is no difference in price because it appears that 10% + 20% = 30%, thus yielding the same discount in both cases. Yet with a little more thought you will realize that this is not correct, since store A calculates a 10% discount on the original list price and a 20% discount on the lower price, while store B calculates the entire 30% discount on the original price. Now, the question to be answered is, what percentage difference is there between the discounts in store A and store B?

One solution might be to assume the cost of the coat to be $100, calculate the 10% discount yielding a $90 price, and take an additional 20% off the $90 price (or $18) to bring the price down to $72, resulting in a 28% discount. In store B, the 30% discount on $100 would bring the price down to $70, giving a discount difference of $2, which in this case means a difference of 2%. This procedure, although correct and not too difficult, is a bit cumbersome and does not give you a full insight into the situation.

An interesting and quite unusual procedure is provided for entertainment and a fresh look into this problem. This method is for obtaining a single percentage discount (or increase) equivalent to two (or more) successive discounts (or increases).

(1) Change each of the percents involved into decimal form:

$$.20 \text{ and } .10.$$

(2) Subtract each of these decimals from 1.00:

$$.80 \text{ and } .90 \text{ (for an increase, add to 1.00)}.$$

(3) Multiply these differences:

$$(.80)(.90) = .72.$$

(4) Subtract this number (i.e., .72) from 1.00:

$1.00 - .72 = .28$, which represents the combined *discount*.

(If the result of step 3 is greater than 1.00, subtract 1.00 from it to obtain the percent of *increase*.)

When we convert .28 back to percent form, we obtain 28%, the equivalent of successive discounts of 20% and 10%.

This combined percentage of 28% differs from 30% by 2%.

The ambitious reader might want to see how this works algebraically. Consider a price p and discounts of $d_1\%$ and $d_2\%$. We have $p-\frac{pd_1}{100}=p\left(1-\frac{d_1}{100}\right)$ representing the price after one discount has been computed. Further, $\left[p\left(1-\frac{d_1}{100}\right)\right]-\left[p\left(1-\frac{d_1}{100}\right)\right]\left(\frac{d_2}{100}\right)=p\left(1-\frac{d_1}{100}\right)\left(1-\frac{d_2}{100}\right)$ represents the price after the second discount has been computed. We should note that $\left(1-\frac{d_1}{100}\right)\left(1-\frac{d_2}{100}\right)$ represents the percent that the new price is of the original price. Therefore, $1-\left(1-\frac{d_1}{100}\right)\left(1-\frac{d_2}{100}\right)$ represents the discount taken off the original price in order to obtain the new price. Thus, the successive discounts are equivalent to a single discount percent of $1-\left(1-\frac{d_1}{100}\right)\left(1-\frac{d_2}{100}\right)$. By translating this final algebraic expression into verbal form, one should be able to establish the technique for converting successive discounts into a single equivalent discount.

Following the same procedure, you can also combine more than two successive discounts. Furthermore, successive increases, combined or not combined with discounts, can also be accommodated in this procedure by *adding* the decimal equivalent of the increase to 1.00 (whereas the discount was subtracted from 1.00) and then continuing the procedure in the same way. If the end result is greater than 1.00, then this end result reflects an overall increase rather than a discount, as found in the above problem.

This procedure not only streamlines a typically cumbersome situation but also provides a glimpse at the overall picture of percentages. For example, the question "Is it advantageous to the buyer in the above problem to receive a 20% discount and then a 10% discount, or the reverse, a 10% discount followed by a 20% discount?" The answer to this question is not intuitively obvious. Yet, since the procedure we just presented shows that the calculation is merely multiplication, a commutative operation, we conclude immediately that there is no difference between the two.

A rather tricky situation occurs when a store offers a 10% discount one day and the next day chooses to increase all prices by 10%. By what percentage have the prices of their products changed? Using our procedure, we come up with a rather surprising result. The first step is to change the discounts and increases to decimal form and either subtract 1.00 or add to 1.00. For the situation described, we have $.90 \times 1.10 = .99$. Since the result is less than 1.00, we have a discount of .01 or 1%. It typically surprises the average consumer that the two percentages do not bring the result to the starting point.

In Figure 2-12, we show what the result of successive discounts and increases bring. As we noted earlier, the results are not intuitively obvious.

Successive changes						
% Discount	20	1.5	10	5	1	.5
& Increase	20	1.5	10	5	1	.5
% Change	.4	2.25	1	.25	.01	.0025

Figure 2-12

One problem that often arises is finding the combinations of successive discounts and increases that will leave the original price unaltered. One possible approach would be to use the conversion technique for a successive discount of *d*% and an increase of *i*% as follows:

$$1-\left(1-\frac{d}{100}\right)\left(1+\frac{i}{100}\right)=0$$

$$1-\left(1-\frac{d}{100}+\frac{i}{100}-\frac{di}{100^2}\right)=0$$

$$100d-100i+di=0$$

$$d=\frac{100i}{100+1}, \text{ or } i=\frac{100d}{100-d}.$$

Figure 2-13 shows some possible values for *d* and *i* that result in zero change.

d	0	9.0909	16.666	20	50	75	10	25
i	0	10	20	25	100	300	11.111	33.333

Figure 2-13

So here you have a delightful algorithm for combining successive discounts, or increases, or combinations of both. Not only is this procedure useful, but it also gives you some newfound power in dealing with percentages when a calculator is not available.

Deceptive Percent Calculations

Another common challenge of working with percents can be seen in a recently reported construction project. A parking lot is going to plant some shrubbery, which will require making each parking space a bit smaller. The spaces will be reduced by 4% in length, and by 5% in width. It is calculated that the area of each space will be reduced by $4 \times 5 = 20\%$. This is a mistake! Consider the following, where a represents the length of the current parking space, and b represents its width. We begin by representing the area of a current parking space as $A_{\text{original}} = a_{\text{original}} \times b_{\text{original}}$. Then the area of new parking space would be

$$\begin{aligned} A_{\text{new}} &= a_{\text{new}} \times b_{\text{new}} = \left(a_{\text{original}} - \frac{4}{100} \times a_{\text{original}}\right) \cdot \left(b_{\text{original}} - \frac{5}{100} \times b_{\text{original}}\right) \\ &= \frac{96}{100} \times a_{\text{original}} \times \frac{95}{100} \times b_{\text{original}} = \frac{114}{125} \times a_{\text{original}} \times b_{\text{original}} \\ &= 0.912 \times a_{\text{original}} \times b_{\text{original}} \approx 0.91 \times A_{\text{original}}. \end{aligned}$$

This shows us that the mistake was quite substantial in that the spaces were actually reduced by about 9%, and not 20%!

By looking at the general case, we can get a better insight into the situation we just encountered. Let length a and width b change by $x\%$ and $y\%$, which can be expressed algebraically as

$$a\left(1 + \frac{x}{100}\right) \cdot b\left(1 + \frac{y}{100}\right) = ab\left(1 + \frac{x+y}{100} + \frac{xy}{10{,}1000}\right).$$

If x and y are numbers where $xy < 100$, then the area will grow between $(x + y)\%$ and $(x + y + 1)\%$. Therefore, the rule-of-thumb is that for percentages $0 < x\%$ and $y\% < 10\%$, the growth of the area is nearly $(x + y)\%$.

This sort of mistake can happen in an even more dramatic fashion. Consider the homeowner who wants to double the volume of his swimming pool (that is, doubling the amount of water it can contain) and decides to simply double each of the dimensions of the pool.

This is a very costly mistake. If we let a, b, and c represent the three dimensions of the pool (width, length, and depth), we determine the pool's volume as $V_{\text{original}} = a_{\text{original}} \times b_{\text{original}} \times c_{\text{original}}$. Then the enlarged pool would be:

$$V_{\text{new}} = a_{\text{new}} \times b_{\text{new}} \times c_{\text{new}} = 2a_{\text{original}} \times 2b_{\text{original}} \times 2c_{\text{original}}$$
$$= 8 \times a_{\text{original}} \times b_{\text{original}} \times c_{\text{original}} = 8V_{\text{original}}.$$

To this homeowner's surprise, he did not double the volume of his pool, but rather increased it by eight times the original volume. One might call this a rather significant mistake!

The Famous Rule of 72

The famous "Rule of 72" states that—roughly speaking—*when money is invested at an annually compounded interest rate of r%, the money will double in $\frac{72}{r}$ years.* So, for example, if we invest money at an 8% compounded annual interest rate, it will double its value in $\frac{72}{8} = 9$ years.

To investigate why, or if, this really works, we consider the compound interest formula:

$$A = P\left(1 + \frac{r}{100}\right)^n,$$

where A is the resulting amount of money, P is the principal invested for n interest periods at r% annually. We need to investigate what happens when $A = 2P$. When we apply the above equation, we get the following:

$$2 = \left(1 + \frac{r}{100}\right)^n. \qquad (1)$$

It then follows that

$$n = \frac{\log 2}{\log\left(1 + \frac{r}{100}\right)}. \qquad (2)$$

Let us make a table of (rounded) values from the above equation, shown in Figure 2-14:

r	n	$n \times r$
1	69.66071689	69.66071689
3	23.44977225	70.34931675
5	14.20669908	71.03349541
7	10.24476835	71.71337846
9	8.043231727	72.38908554
11	6.641884618	73.0607308
13	5.671417169	73.72842319
15	4.959484455	74.39226682

Figure 2-14

The arithmetic mean (the average) of the nr values is 72.04092673, which is quite close to 72, and so our "Rule of 72" seems to be an accurate estimate for doubling money at an annual interest rate of $r\%$ for n interest periods.

An ambitious reader might try to determine a "rule" for tripling and quadrupling money in a manner similar to the way we dealt with the doubling of money. The above equation (2) for k-tupling would be

$$n = \frac{\log k}{\log\left(1 + \frac{r}{100}\right)},$$

which for $r = 8$ gives the value for $n = 29.91884022 \times (\log k)$. Thus $nr = 239.3507218 \times \log k$, which for $k = 3$ (the tripling effect) gives us $nr = 114.1821673$. We could then say that for tripling money we would have a "rule of 114."

Extracting a Square Root

Why would anyone want to find the square root of a number without using a calculator? Surely, no one would do such a thing. However, it is possible that there are some people who are curious to know what is actually being done to find a square root. They could develop a better understanding of the process and briefly gain some independence from the calculator. We will use a method that's generally not taught in school, but gives valuable insight into the meaning of a square root. The beauty of this method is that it really allows you to understand what is going on. This method was first published by the English mathematician Joseph Raphson (1668–1715) in his 1690 book *Analysis alquationum universalis.* Raphson attributed the method to Isaac Newton, and therefore, the algorithm bears both names as the *Newton-Raphson method.*

It is perhaps best to see the method used in a specific example. Suppose we wish to find $\sqrt{27}$. Obviously, the calculator would be used here. However, you might like to guess what this value might be. Certainly, it is between $\sqrt{25}$ and $\sqrt{36}$, or between 5 and 6, but closer to 5.

Suppose we guess at 5.2. If this were the correct square root, then if we were to divide 27 by 5.2, we would get 5.2. But this is not the case here, since $\sqrt{27} \neq 5.2$.

We seek a closer approximation. To do that, we find $\frac{27}{5.2} = 5.192$. Since $27 \approx 5.2 \times 5.192$, one of the factors (5.2 in this case) must be greater than $\sqrt{27}$ and the other factor (5.192 in this case) must be less than $\sqrt{27}$. Hence, $\sqrt{27}$ is sandwiched between the two numbers 5.2 and 5.192, that is, $5.192 < \sqrt{27} < 5.2$. It is plausible that the average of 5.2 and 5.192 (5.196) is a better approximation for $\sqrt{27}$ than either number.

This process continues, each time with additional decimal places, so that an allowance is made for a closer approximation. That is,

$$\frac{5.192+5.196}{2} = 5.194,$$

then

$$\frac{27}{5.194} = 5.19831.$$

This continuous process provides insight into the finding of the square root of a number that is not a perfect square.

Cumbersome as the method may be, it helps you understand what a square root represents. Why this neat method wasn't taught in the schools in the days before the calculator remains a mystery.

Are Averages in Fact Averages?

Although baseball batting averages permeate sports discussions, few people realize that these "baseball batting averages" aren't really "averages" in the true sense of the term. They are actually percentages. Most people, especially after trying to explain this concept, will begin to realize that it is not an average in the way they usually define an "average": as the arithmetic mean. It might be illustrative to search the sports section of your local newspaper for two baseball players who currently have the same batting average but who have achieved their respective averages with a different number of hits. We shall use a hypothetical example here.

Consider two players, David and Lisa, each with a batting average of .667. David achieved his batting average by getting 20 hits for 30 at-bats, while Lisa achieved her batting average by getting 2 hits for 3 at-bats.

In the next game both players performed equally, getting 1 hit for 2 at-bats for a .500 batting average. One might expect that David and Lisa would still have the same batting average at the end of the day. Calculating their respective averages:

David now has $20 + 1 = 21$ hits for $30 + 2 = 32$ at-bats for a $\frac{21}{32} = .656$ batting average.

Lisa now has $2 + 1 = 3$ hits for $3 + 2 = 5$ at-bats for a $\frac{3}{5} = .600$ batting average. Surprise! They do not have equal batting averages.

Suppose we consider the next game, where Lisa performs considerably better than David. Lisa gets 2 hits for 3 at-bats, while David gets 1 hit for 3 at-bats. We shall now calculate their respective averages:

David has $21 + 1 = 22$ hits for $32 + 3 = 35$ at-bats for a batting average of $\frac{22}{35} = .629$.

Lisa has $3 + 2 = 5$ hits for $5 + 3 = 8$ at-bats for a batting average of $\frac{5}{8} = .625$.

Amazingly, despite Lisa's superior performance on this day, her batting average, which was the same as David's at the start, is still lower.

There is much to be learned from this "misuse" of the word "average," but more importantly, one can appreciate the notion of how averages are affected by the varying weights of items being averaged.

Howlers

In the early years of schooling, we learned to reduce fractions whenever possible. There were specific ways to do this correctly. Some wise guy seems to have come up with a shorter way to reduce some fractions. Is he right? He was asked to reduce the fraction $\frac{26}{65}$ and did it in the following way:

$$\frac{2\not{6}}{\not{6}5}=\frac{2}{5}.$$

That is, he just canceled out the 6s to get the right answer. Is this procedure correct? Can it be extended to other fractions? If so, then we were surely treated unfairly by our elementary school teachers, who made us do much more work. Let's look at what was done here and if it can be generalized.

In his book *Fallacies in Mathematics*, the Scottish mathematician E.A. Maxwell (1907–1987) refers to the following cancellations as "howlers":

$$\frac{1\not{6}}{\not{6}4}=\frac{1}{4} \quad \frac{2\not{6}}{\not{6}5}=\frac{2}{5}.$$

Perhaps when someone did the fraction reductions this way and still got the right answer, it just made them howl. Begin by reducing the following fractions to lowest terms:

$$\frac{16}{64}, \ \frac{19}{95}, \ \frac{26}{65}, \ \frac{49}{98}.$$

After you have reduced to lowest terms each of the fractions in the usual manner, one may ask why it couldn't have been done in the following way:

$$\frac{1\not{6}}{\not{6}4}=\frac{1}{4}$$

$$\frac{1\not{9}}{\not{9}5}=\frac{1}{5}$$

$$\frac{2\not{6}}{\not{6}5}=\frac{2}{5}$$

$$\frac{4\not{9}}{\not{9}8}=\frac{4}{8}=\frac{1}{2}.$$

At this point, you may be amazed. Your first reaction is probably to ask if this can be done to any fraction composed of two-digit numbers.

Can you find another fraction (comprised of two-digit numbers) where this type of cancellation will work? You might cite

$$\frac{5\not{5}}{\not{5}5}=\frac{5}{5}=1$$

as an illustration of this type of cancellation. This will hold true for all two-digit multiples of eleven, which is trivial, and our concern is only with proper fractions (i.e., whose value is less than one).

For readers with a good working knowledge of elementary algebra, we can "explain" this awkward situation. The four fractions above are the *only* fractions (composed of two-digit numbers) where this type of cancellation holds true.

An arithmetic explanation for why this works can be seen from the following calculation:

$$\frac{16}{64}=\frac{1\times 10+6}{10\times 6+4}=\frac{\not{6}\times\frac{16}{6}}{\not{6}\times\frac{64}{6}}=\frac{\not{6}\times\frac{8}{3}}{\not{6}\times\frac{32}{3}}=\frac{8}{32}=\frac{1}{4},$$

therefore,

$$\frac{16}{64}=\frac{1\times 10+\not{6}}{10\times\not{6}+4}=\frac{1}{4}.$$

Consider the fraction

$$\frac{10x+a}{10a+y}.$$

The above four fractions were such that by canceling the a's the fraction was equal to $\frac{x}{y}$. Therefore,

$$\frac{10x+a}{10a+y}=\frac{x}{y}.$$

This yields:

$$\begin{aligned}
&y(10x+a)=x(10a+y)\\
&10xy+ay=10ax+xy\\
&9xy+ay=10ax.
\end{aligned}$$

And so

$$y=\frac{10ax}{9x+a}.$$

At this point we shall inspect this equation. It is necessary that x, y, and a are integers, since they were digits in the numerator and denominator of a fraction. It is now our task to find the values of a and x for which y will also be an integer.

To avoid a lot of algebraic manipulation, you may want to set up a chart for generating integer values of y from

$$y=\frac{10ax}{9x+a}.$$

Remember that x, y, and a must be single-digit integers. Figure 2-15 presents a portion of the table you can construct. Notice that the cases where $x=a$ are excluded since $\frac{x}{a}=1$.

x/a	**1**	**2**	**3**	**4**	**5**	**6**	**7**	**8**	**9**
1		$\frac{20}{11}$	$\frac{30}{12}$	$\frac{40}{13}$	$\frac{50}{14}$	$\frac{60}{15}=\mathbf{4}$	$\frac{70}{16}$	$\frac{80}{17}$	$\frac{90}{18}=\mathbf{5}$
2	$\frac{20}{19}$		$\frac{60}{21}$	$\frac{80}{22}$	$\frac{100}{23}$	$\frac{120}{24}=\mathbf{5}$	$\frac{140}{25}$	$\frac{160}{26}$	$\frac{180}{27}$
3	$\frac{30}{28}$	$\frac{60}{29}$		$\frac{120}{31}$	$\frac{150}{32}$	$\frac{180}{33}$	$\frac{210}{34}$	$\frac{240}{35}$	$\frac{270}{36}$
4	$\frac{40}{37}$	$\frac{80}{38}$	$\frac{120}{39}$		$\frac{200}{41}$	$\frac{240}{42}$	$\frac{280}{43}$	$\frac{320}{44}$	$\frac{360}{45}=\mathbf{8}$
5	$\frac{50}{46}$	$\frac{100}{47}$	$\frac{150}{48}$	$\frac{200}{49}$		$\frac{300}{51}$	$\frac{350}{52}$	$\frac{400}{53}$	$\frac{450}{54}$
6	$\frac{60}{55}$	$\frac{120}{56}$	$\frac{180}{57}$	$\frac{240}{58}$	$\frac{300}{59}$		$\frac{420}{61}$	$\frac{480}{62}$	$\frac{540}{63}$
7	$\frac{70}{64}$	$\frac{140}{65}$	$\frac{210}{66}$	$\frac{280}{67}$	$\frac{350}{68}$	$\frac{420}{69}$		$\frac{560}{71}$	$\frac{630}{72}$
8	$\frac{80}{73}$	$\frac{160}{74}$	$\frac{240}{75}$	$\frac{320}{76}$	$\frac{400}{77}$	$\frac{480}{78}$	$\frac{560}{79}$		$\frac{720}{81}$
9	$\frac{90}{82}$	$\frac{180}{83}$	$\frac{270}{84}$	$\frac{360}{85}$	$\frac{450}{86}$	$\frac{540}{87}$	$\frac{630}{88}$	$\frac{720}{89}$	

Figure 2-15

The portion of the chart pictured above shows four integer values of y; that is, when $x=1, a=6$, then $y=4$, and when $x=2, a=6$, then $y=5$. These values yield the fractions $\frac{16}{24}$ and $\frac{26}{65}$, respectively. The remaining two integral values of y are obtained when $x=1$, and $a=9$, yielding $y=5$, and when $x=4$ and $a=9$, yielding $y=8$. These yield the fractions $\frac{19}{95}$ and $\frac{49}{98}$, respectively. This should convince you that there are only four such "howling" fractions composed of two-digit numbers.

You may now wonder if there are fractions composed of numerators and denominators with more than two digits where this strange type of cancellation holds true.

What follows are some examples of three-digit numbers where this sort of strange cancellation can work.

$$\frac{199}{995}=\frac{1\cancel{9}\cancel{9}}{\cancel{9}\cancel{9}5}\left(=\frac{1}{5}\right),\quad \frac{266}{665}=\frac{2\cancel{6}\cancel{6}}{\cancel{6}\cancel{6}5}\left(=\frac{2}{5}\right),\quad \frac{124}{217}=\frac{\cancel{1}\cancel{2}4}{\cancel{2}\cancel{1}7}\left(=\frac{4}{7}\right),$$

$$\frac{103}{206}=\frac{1\cancel{0}3}{2\cancel{0}6}=\frac{13}{26}\left(=\frac{1}{2}\right),\quad \frac{495}{990}=\frac{4\cancel{9}5}{9\cancel{9}0}=\frac{45}{90}\left(=\frac{1}{2}\right),$$

$$\frac{165}{660}=\frac{1\cancel{6}5}{6\cancel{6}0}=\frac{15}{60}\left(=\frac{1}{4}\right),\quad \frac{127}{762}=\frac{1\cancel{2}\cancel{7}}{\cancel{7}6\cancel{2}}\left(=\frac{1}{6}\right), \text{ and}$$

$$\frac{143185}{1701856}=\frac{143\cancel{1}\cancel{8}5}{170\cancel{1}\cancel{8}56}=\frac{1435}{17056}\left(=\frac{35}{416}\right).$$

Try this type of cancellation with $\frac{499}{998}$. You should find that

$$\frac{4\cancel{9}\cancel{9}}{\cancel{9}\cancel{9}8}=\frac{4}{8}=\frac{1}{2}.$$

A pattern is now emerging, and you may realize that

$$\frac{49}{98}=\frac{499}{998}=\frac{4999}{9998}=\frac{49999}{99998}=\ldots.$$

$$\frac{16}{64}=\frac{166}{664}=\frac{1666}{6664}=\frac{16666}{66664}=\frac{166666}{666664}=\ldots.$$

$$\frac{19}{95}=\frac{199}{995}=\frac{1999}{9995}=\frac{19999}{99995}=\frac{199999}{999995}=\ldots.$$

$$\frac{26}{65}=\frac{266}{665}=\frac{2666}{6665}=\frac{26666}{66665}=\frac{266666}{666665}=\ldots.$$

Enthusiastic readers may wish to justify these extensions of the original howlers. Readers who, at this point, desire to seek out additional fractions that permit this strange cancellation should consider the following fractions. You may wish to verify the legitimacy of this strange cancellation and then set out to discover more such fractions.

$$\frac{3\cancel{3}2}{8\cancel{3}0}=\frac{32}{80}=\frac{2}{5}$$

$$\frac{3\cancel{8}5}{8\cancel{8}0}=\frac{35}{80}=\frac{7}{16}$$

$$\frac{1\cancel{3}8}{\cancel{3}45}=\frac{18}{45}=\frac{2}{5}$$

$$\frac{2\cancel{7}5}{7\cancel{7}0}=\frac{25}{70}=\frac{5}{14}$$

$$\frac{1\cancel{6}\cancel{3}}{\cancel{3}2\cancel{6}}=\frac{1}{2}.$$

However, be careful, because

$$\frac{16\cancel{3}}{\cancel{3}26}\neq\frac{1}{2}\text{ and }\frac{1\cancel{6}3}{32\cancel{6}}\neq\frac{1}{2}.$$

Aside from providing an algebraic application, which can be used to introduce a number of important topics in a motivational way, this topic can also provide some recreational activities. Here are some more "howlers."

$$\frac{4\cancel{8}\cancel{4}}{\cancel{8}\cancel{4}7}=\frac{4}{7}\qquad\frac{\cancel{5}\cancel{4}5}{6\cancel{5}\cancel{4}}=\frac{5}{6}\qquad\frac{\cancel{4}\cancel{2}4}{7\cancel{4}\cancel{2}}=\frac{4}{7}\qquad\frac{24\cancel{9}}{\cancel{9}96}=\frac{24}{96}=\frac{1}{4}$$

$$\frac{4\cancel{8}\cancel{4}\cancel{8}\cancel{4}}{\cancel{8}\cancel{4}\cancel{8}\cancel{4}7}=\frac{4}{7}\qquad\frac{\cancel{5}\cancel{4}\cancel{5}\cancel{4}5}{6\cancel{5}\cancel{4}\cancel{5}\cancel{4}}=\frac{5}{6}\qquad\frac{\cancel{4}\cancel{2}\cancel{4}\cancel{2}4}{7\cancel{4}\cancel{2}\cancel{4}\cancel{2}}=\frac{4}{7}$$

$$\frac{\cancel{3}\cancel{2}\cancel{4}3}{4\cancel{3}\cancel{2}\cancel{4}}=\frac{3}{4}\qquad\frac{\cancel{6}\cancel{4}\cancel{8}6}{8\cancel{6}\cancel{4}\cancel{8}}=\frac{6}{8}=\frac{3}{4}$$

$$\frac{147\!\!\!/\,1\!\!\!/\,4\!\!\!/}{7\!\!\!/\,1\!\!\!/\,4\!\!\!/\,68}=\frac{14}{68}=\frac{7}{34} \qquad \frac{\cancel{8}\cancel{7}\cancel{8}\cancel{0}\cancel{4}8}{9\cancel{8}\cancel{7}\cancel{8}\cancel{0}\cancel{4}}=\frac{8}{9}$$

$$\frac{1\cancel{4}\cancel{2}\cancel{8}\cancel{5}\cancel{7}\cancel{1}}{\cancel{4}\cancel{2}\cancel{8}\cancel{5}\cancel{7}\cancel{1}3}=\frac{1}{3} \qquad \frac{2\cancel{8}\cancel{5}\cancel{7}\cancel{1}\cancel{4}\cancel{2}}{\cancel{8}\cancel{5}\cancel{7}\cancel{1}\cancel{4}\cancel{2}6}=\frac{2}{6}=\frac{1}{3} \qquad \frac{3\cancel{4}\cancel{6}\cancel{1}\cancel{5}\cancel{3}\cancel{8}}{\cancel{4}\cancel{6}\cancel{1}\cancel{5}\cancel{3}\cancel{8}4}=\frac{3}{4}$$

$$\frac{\cancel{7}\cancel{6}\cancel{7}\cancel{1}\cancel{2}\cancel{3}\cancel{2}\cancel{8}7}{8\cancel{7}\cancel{6}\cancel{7}\cancel{1}\cancel{2}\cancel{3}\cancel{2}\cancel{8}}=\frac{7}{8} \qquad \frac{\cancel{3}\cancel{2}\cancel{4}\cancel{3}\cancel{2}\cancel{4}\cancel{3}\cancel{2}\cancel{4}3}{4\cancel{3}\cancel{2}\cancel{4}\cancel{3}\cancel{2}\cancel{4}\cancel{3}\cancel{2}\cancel{4}}=\frac{3}{4}$$

$$\frac{\cancel{1}\cancel{0}\cancel{2}\cancel{5}\cancel{6}\cancel{4}1}{4\cancel{1}\cancel{0}\cancel{2}\cancel{5}\cancel{6}\cancel{4}}=\frac{1}{4} \qquad \frac{\cancel{3}\cancel{2}\cancel{4}\cancel{3}\cancel{2}\cancel{4}3}{4\cancel{3}\cancel{2}\cancel{4}\cancel{3}\cancel{2}\cancel{4}}=\frac{3}{4} \qquad \frac{4\cancel{5}\cancel{7}\cancel{1}\cancel{4}\cancel{2}\cancel{8}}{\cancel{5}\cancel{7}\cancel{1}\cancel{4}\cancel{2}\cancel{8}5}=\frac{4}{5}$$

$$\frac{4\cancel{8}\cancel{4}\cancel{8}\cancel{4}\cancel{8}\cancel{4}}{\cancel{8}\cancel{4}\cancel{8}\cancel{4}\cancel{8}\cancel{4}7}=\frac{4}{7} \qquad \frac{5\cancel{9}\cancel{5}\cancel{2}\cancel{3}\cancel{8}\cancel{0}}{\cancel{9}\cancel{5}\cancel{2}\cancel{3}\cancel{8}\cancel{0}8}=\frac{5}{8} \qquad \frac{\cancel{4}\cancel{2}\cancel{8}\cancel{5}\cancel{7}\cancel{1}4}{6\cancel{4}\cancel{2}\cancel{8}\cancel{5}\cancel{7}\cancel{1}}=\frac{4}{6}=\frac{2}{3}$$

$$\frac{\cancel{5}\cancel{4}\cancel{5}\cancel{4}\cancel{5}\cancel{4}5}{6\cancel{5}\cancel{4}\cancel{5}\cancel{4}\cancel{5}\cancel{4}}=\frac{5}{6} \qquad \frac{6\cancel{9}\cancel{2}\cancel{3}\cancel{0}\cancel{7}\cancel{6}}{\cancel{9}\cancel{2}\cancel{3}\cancel{0}\cancel{7}\cancel{6}8}=\frac{6}{8}=\frac{3}{4} \qquad \frac{\cancel{4}\cancel{2}\cancel{4}\cancel{2}\cancel{4}\cancel{2}4}{7\cancel{4}\cancel{2}\cancel{4}\cancel{2}\cancel{4}\cancel{2}}=\frac{4}{7}$$

$$\frac{\cancel{5}\cancel{3}\cancel{8}\cancel{4}\cancel{6}\cancel{1}5}{7\cancel{5}\cancel{3}\cancel{8}\cancel{4}\cancel{6}\cancel{1}}=\frac{5}{7} \qquad \frac{\cancel{2}\cancel{0}\cancel{5}\cancel{1}\cancel{2}\cancel{8}2}{8\cancel{2}\cancel{0}\cancel{5}\cancel{1}\cancel{2}\cancel{8}}=\frac{2}{8}=\frac{1}{4} \qquad \frac{\cancel{3}\cancel{1}\cancel{1}\cancel{6}\cancel{8}\cancel{8}3}{8\cancel{3}\cancel{1}\cancel{1}\cancel{6}\cancel{8}\cancel{8}}=\frac{3}{8}$$

$$\frac{\cancel{6}\cancel{4}\cancel{8}\cancel{6}\cancel{4}\cancel{8}6}{8\cancel{6}\cancel{4}\cancel{8}\cancel{6}\cancel{4}\cancel{8}}=\frac{6}{8}=\frac{3}{4} \qquad \frac{4\cancel{8}\cancel{4}\cancel{8}\cancel{4}\cancel{8}\cancel{4}\cancel{8}\cancel{4}}{\cancel{8}\cancel{4}\cancel{8}\cancel{4}\cancel{8}\cancel{4}\cancel{8}\cancel{4}7}=\frac{4}{7}$$

Mathematics continues to hold hidden treasures wrapped up within "mistaken procedures."

Rounding-Off Can Cause a Mistaken Answer

There are times when rounding off correctly leads to a wrong answer. Consider the following example. At an airport there are 963 stranded passengers. Buses are ordered to take these passengers away. Each bus can hold 59 passengers. The question is: how many buses are required to take these passengers to their destination? Typically, a student would do the following calculation: $963 \div 59 \approx 16.32203389$. Since the number of buses must be a whole number, the student would round off the answer to 16, since the 3 following the decimal point is less than 5. Although the calculation and rounding-down were correct, this answer clearly does not solve the problem. Obviously, 17 buses will be required, and the seventeenth bus will not be completely full. This is an example of an incorrect answer in spite of correct calculations.

Stumbling Blocks: The Nemesis of the Zero

Let's inspect the following arithmetic: $\frac{0}{3}=0$, and $\frac{0}{5}=0$; therefore, we might conclude that $\frac{0}{0}=0$, since the numerator is zero. Since $\frac{3}{3}=1$, and $\frac{5}{5}=1$, we might conclude that $\frac{0}{0}=1$, since the numerator and the denominator are equal. Because of this dilemma, we call this fraction $\frac{0}{0}$ *indeterminate*, and it has no value.

Why does the zero play such a confusing role in arithmetic? In addition, zero plays an absolutely neutral role, as can be seen in the following example: $a+0=0+a=a$ for all (real) numbers a. However, zero creates a stumbling block with subtraction. There is a difference between subtracting zero from a number and subtracting a number from zero: $a - 0 = a$, whereas $0 - a = -a$ for all (real) numbers a.

The situation gets more complicated for multiplication, and even more so when we look at division involving zero. In the case of multi plication, we see that zero seems to swallow up anything with which it is multiplied; in other words, zero times anything is zero: $a\times 0=0\times a=0$ for all (real) numbers a.

Another stumbling block is the confusion about the roles of zero and 1 in multiplication and addition. We know that when we multiply any number by 1, the number stays the same. Analogously, when we add zero to any number, that number also remains the same, as we can see from the following: $a\times 1=1\times a=a$, and $a+0=0+a=a$ for all (real) numbers a. Does that imply that $1 = 0$? Certainly not!

The situation gets more dramatic when we consider division involving zero. It makes a big difference whether zero appears in the numerator or the denominator. If we divide zero by any number other than zero, we get zero, such as $0 \div a = \frac{0}{a} = 0$ for all (real) numbers a.

However, if we divide any number by zero, we end up with an inexplicable situation. Thus, we say division by zero is not allowed. Many mathematics teachers instruct their students that the "the eleventh commandment" is thou shall not divide by zero! Violation of this commandment will become clear in the following discussion. So, brace yourself for some magnificent mistakes.

There are many ways in which we can "prove" that $1 = 2$. Now, one wonders, what does this mean? How can one prove that $1 = 2$? Clearly, there must be an error somewhere in the "proof." What's more, if this error is well hidden, then our "proof" becomes even more frustrating. Yet, knowing that $1 = 2$ cannot be true, we search on, in the hope of locating this mistake. We shall begin with a simple mistake, one that can be seen from various points of view.

Violating the "Eleventh Commandment"— A Mistake!

We can "prove" that $1 = 2$ by using the forbidden division by zero. We know that if $5a = 5b$, then $a = b$. However, if we use the reasoning that $1 \times 0 = 0$, and $2 \times 0 = 0$, we can conclude that $1 \times 0 = 2 \times 0$. Then, using the same reasoning as before, we can say that $1 = 2$ by dividing both sides by zero. Since such absurdities result from division by zero, it is forbidden in mathematics. Remember the eleventh commandment: Thou shall not divide by zero.

Here is a simple example of a violation of this rule:

$$12 - 12 = 18 - 18$$

$$12 - 8 - 4 = 18 - 12 - 6$$

$$2(6 - 4 - 2) = 3(6 - 4 - 2)$$

$$2 = 3.$$

We can even create a comical situation by violating this important rule. Through the following argument, we can show that no one is overweight. In other words, everybody's weight is correct for their body.

Follow this proof carefully. We shall begin by letting G = actual weight, g = ideal weight, W = overweight.

The definition of our actual weight is $G = g + W$.

Subtract g from both sides of the equation to get $G - g = W$.

Multiply both sides by $(G - g)$ to get $(G - g)^2 = W(G - g)$.

Expand the expression to get $G^2 - 2Gg + g^2 = GW - gW$.

Subtract g^2 from both sides of the equation to get $G^2 - 2Gg = GW - g^2 - gW$.

Now subtract GW from both sides of the equation to get $G^2 - 2Gg - GW = -g^2 - gW$.

We will now add Gg to both sides equation to get $G^2 - Gg - GW = Gg - g^2 - gW$.

Simplifying further gives us $G(G - g - W) = g(G - g - W)$.

Dividing both sides by $(G - g - W)$ leaves us with $G = g$, which essentially states that our actual weight is our ideal weight. We then ask ourselves, where does the mistake lurk? Recall that if $G = g + W$, then $G - g - W = 0$. Then take note of the step that led us to the absurd conclusion above. Yes, it was division by zero.

Adding Fractions: Wrong Method, Right Sum!

Suppose we want to add the two fractions

$$\frac{9}{3} + \frac{-16}{4}.$$

By adding the numerators and the denominators—clearly, the wrong procedure—we have

$$\frac{9}{3} + \frac{-16}{4} = \frac{9-16}{3+4},$$

which surprisingly gives us a correct answer:

$$\frac{9-16}{3+4} = \frac{-7}{7} = -1.$$

Remember, of course, that the procedure was certainly wrong!

The correct procedure is

$$\frac{9}{3}+\frac{-16}{4}=\frac{36}{12}-\frac{48}{12}=-\frac{12}{12}=-1.$$

As you can see, this gives us the correct answer, the same as the one before. Thus, a correct answer does not support an incorrect procedure!

We can look at another example of how this faulty procedure curiously leads to a correct answer:

$$\frac{-5}{-1}+\frac{20}{2}=\frac{-5+20}{-1+2}=\frac{15}{1}=15.$$

Done correctly, the calculation would look like this:

$$\frac{-5}{-1}+\frac{20}{2}=5+10=15,$$

which is the same answer. Don't be fooled into believing that adding the numerators and denominators is a simplified way of adding fractions, even though it leads to a correct answer in these particular cases. It is still a mistaken procedure.

A comparison of the wrong method with the correct one could be useful. Consider the following:

The wrong method is

$$\frac{a}{b}+\frac{u}{v}=\frac{a+u}{b+v}.$$

The correct method is

$$\frac{a}{b}+\frac{u}{v}=\frac{av+bu}{bv}.$$

For what values of a, b, u, and v would this fallacious method produce a correct answer?

A Crazy (Mistaken) Procedure to Multiply Fractions Leads to a Correct Answer

Here we are given two binomials consisting of fractions, which we then multiply in a rather strange way:

$$\left(\frac{2}{3}-\frac{5}{4}\right)\cdot\left(\frac{1}{3}+\frac{5}{8}\right)=\frac{2}{3}\cdot\frac{1}{3}-\frac{5}{4}\cdot\frac{5}{8}.$$

This erroneous procedure has us multiplying the first two members of each parenthetical expression and the second two members of each parenthetical expression. Doing the indicated arithmetic, we get the following correct answer:

$$\frac{2}{3}\cdot\frac{1}{3}-\frac{5}{4}\cdot\frac{5}{8}=\frac{2}{9}-\frac{25}{32}=-\frac{161}{288}=-0.55902\overline{7}.$$

Let's compare this to the correct procedure, just to make sure that, in fact, this answer is truly correct:

$$\left(\frac{2}{3}-\frac{5}{4}\right)\cdot\left(\frac{1}{3}+\frac{5}{8}\right)=-\frac{7}{12}\cdot\frac{23}{24}=-\frac{161}{288}=-0.559027.$$

Once again, it was a matter of luck that the correct answer was obtained with a completely mistaken procedure. This should teach us that if something "works" for one case, it is not necessarily generalizable. Again, it would be interesting for the motivated reader to determine under what conditions the mistaken method will result in a correct answer.

Ignoring Fractions Can (Mistakenly) Lead to a Correct Answer

Can "annoying" fractions simply be ignored? Can you imagine that the product

$$\left(a+\frac{b}{c}\right)\cdot\left(x-\frac{y}{z}\right)=a\times x?$$

Is it possible that we can just disregard the two fractions $\frac{b}{c}$ and $\frac{y}{z}$ and still get the correct answer? Let's take a look at the following calculation, where we will do just that—ignore the fractions!

$$\left(7+\frac{3}{7}\right)\cdot\left(4-\frac{3}{13}\right)=7\cdot 4=28.$$

How is this possible? Just look at the calculation.

First, we (correctly) multiply the two binomial expressions, and then simplify:

$$\left(7+\frac{3}{7}\right)\cdot\left(4-\frac{3}{13}\right)=7\times 4-7\times\frac{3}{13}+\frac{3}{7}\times 4-\frac{3}{7}\times\frac{3}{13}=28-\frac{21}{13}+\frac{12}{7}-\frac{9}{91}$$
$$=28-\frac{147}{91}+\frac{156}{91}-\frac{9}{91}=28.$$

Or you can also get the correct answer by converting each parenthetical expression to an improper fraction and then multiplying:

$$\left(\frac{52}{7}\right)\cdot\left(\frac{49}{13}\right)=28.$$

Yes, we arrived at that same answer as we did when we ignored the fraction portions of the binomial expressions.

The same weird result can be obtained with the following example:

$$\left(7+\frac{1}{2}\right)\cdot\left(5-\frac{1}{3}\right)=7\times 5=35,$$

which we can easily verify just as we did above. In case you are still not convinced, here is yet another example of eliminating the "annoying" fractions—making the same mistake—and still getting the correct answer:

$$\left(31+\frac{1}{2}\right)\cdot\left(21-\frac{1}{3}\right)=31\times 21=651.$$

To make this point further, consider another example:

$$\left(6+\frac{1}{4}\right)\cdot\left(5-\frac{1}{5}\right)=6\times5=30.$$

Yes, the above calculations also (strangely) lead to the correct answer. Wow! What is actually going on here? Might this be a delightful simplification for future calculations? However, much to everyone's disappointment, this is not true for all cases of this form. Yes, it would be a mistake to extend this principle to all such mixed numbers. Take, for example, the following:

$$\left(7+\frac{1}{5}\right)\cdot\left(4-\frac{2}{3}\right)=24.$$

If we had used the "rule" above, we would have gotten $7 \times 4 = 28$, which is not equal to the correct answer.

At this point you must be curious to know when the following is true:

$$\left(a+\frac{b}{c}\right)\cdot\left(x-\frac{y}{z}\right)=a\times x.$$

Let's consider the following product, where we can inspect the situation:

$$\left(7+\frac{1}{2}\right)\cdot\left(5-\frac{1}{3}\right)=7\times5=35.$$

Consider each of the two binomials separately: $7+\frac{1}{2} = 15 \div 2$, and $5-\frac{1}{3} = 14 \div 3$.

Notice that 2 is a divisor of 14, and 3 is a divisor of 15. Any non-prime integers differing by a small amount in comparison with the divisors can be used.

Although we are in the chapter on arithmetic, we will once again protrude to the topic of algebra, since it allows us to understand why, and when, this weird mistake can happen. Suppose we are given *a* and

b and are looking for values for u and v so that $(a+u)(b-v)=ab$. This is essentially what was demonstrated with the above examples. To get this result, we must find u and v where $ub-av-uv=0$, which gets us to $v=\frac{bu}{a+u}$, where u can take on any value.

This rather magnificent mistake can also be extended to three terms in an expression, as shown with the following two examples:

$$\left(2-\frac{2}{13}\right)\cdot\left(6+\frac{1}{4}\right)\cdot\left(5+\frac{1}{5}\right)=2\cdot6\cdot5=60$$

$$\left(2-\frac{2}{13}\right)\cdot\left(4+\frac{1}{6}\right)\cdot\left(5+\frac{1}{5}\right)=2\cdot4\cdot5=40.$$

The ambitious reader may want to find other such special products.

We will look at the product of two numbers, each of which is formed by the same digit repeated the same number of times as shown in Figure 2-16. The Arabic book *Condensing Arithmetic Operations*, written around the year 1300, provided the following calculations:

Number of 1s	The equal factors		Product
1	1 × 1	=	1
2	11 × 11	=	121
3	111 × 111	=	12321
4	1,111 × 1,111	=	1234321
...	...		...
9	111,111,111 × 111,111,111	=	12345678987654321

Figure 2-16

One would expect that this pattern of palindromic numbers would continue. However, as shown in Figure 2-17, the number generated by squaring the number with ten ones destroys the pattern. Thus, to generalize here would be to once again make an arithmetic mistake.

Number of 1s	The equal factors		Products
1	1×1	=	1
2	11×11	=	121
3	111×111	=	12321
4	$1{,}111 \times 1{,}111$	=	1234321
5	$11{,}111 \times 11{,}111$	=	123454321
6	$111{,}111 \times 111{,}111$	=	12345654321
7	$1{,}111{,}111 \times 1{,}111{,}111$	=	1234567654321
8	$11{,}111{,}111 \times 11{,}111{,}111$	=	123456787654321
9	$111{,}111{,}111 \times 111{,}111{,}111$		12345678987654321
10	$1{,}111{,}111{,}111 \times 1{,}111{,}111{,}111$	=	

Figure 2-17

You might try the calculations to see why this pattern doesn't continue as you might have expected.

Strange Exponential Relationships That Can Lead to a Mistaken Process

Consider the following calculation: $2^5 \times 9^2 = 32 \times 81 = 2592$. Notice how the digits in the two bases (2, 9) and the two exponents (5, 2) line up in order within the product 2592. Can we derive a generalized rule from this example? Or would it be a mistake? Much to our disappointment, this is the only time such a pattern exists. This misleading relationship was first discovered by English mathematician Henry E. Dudeney (1857–1930). It is quite easy to find a counterexample: $2^2 \times 2^2 - 16$, and not 2222. In 1934, the recreational mathematics proponent Charles W. Trigg[1] showed that the only possible solution for such a relationship is the above example.

With a different process, this time subtraction, we have $8^2 - 2^2 = 82 - 22 = 60$, which turns out to be correct, since $64 - 4 = 60$. Also, $9^2 - 1^2 = 92 - 12 = 80$, which, again, is correct, since $81 - 1 = 80$. But please do not make the mistake by applying this method to other such situations.

[1] Robinson, Raphael. "C. W. Trigg: E 69." *The American Mathematical Monthly* 41, no. 5 (1934), p. 332.

To further protect the reader from making mistakes by generalizing a lovely pattern, look at the following "rule":

"If we seek the square of the sum of the digits of a number, just drop the plus signs and the exponent."

There are other weird arrangements in arithmetic that are entertaining but cautionary because the arrangement is possible only in certain situations. As an example, notice the symmetry of the of the following calculation: $(8+1)^2 = 81$. This can also hold for cubing a number, as with the following example: $(5+1+2)^3 = 512$ $(= 8^3)$. Figure 2-18 provides a list of other such examples where dropping the plus sign happens to yield the correct answer.

(Sum of the Digits)^n		Power		Number
4^1	=	4^1	=	4
$(8 + 1)^2$	=	9^2	=	81
$(5+1+2)^3$	=	8^3	=	512
$(1+9+6+8+3)^3$	=	27^3	=	19,683
$(2+4+0+1)^4$	=	7^4	=	2,401
$(1+6+7+9+6+1+6)^4$	=	36^4	=	1,679,616
$(5+2+5+2+1+8+7+5)^5$	=	$\mathbf{35^5}$	=	52,521,875
$(2+0+5+9+6+2+9+7+6)^5$	=	46^5	=	205,962,976
$(3+4+0+1+2+2+2+4)^6$	=	18^6	=	34,012,224
$(2+4+7+9+4+9+1+1+2+9+6)^6$	=	54^6	=	24,794,911,296
$(6+1+2+2+2+0+0+3+2)^7$	=	18^7	=	612,220,032

Figure 2-18

Unfortunately, although this procedure is nifty, it would be a huge mistake to generalize it. It is sometimes tempting to search for reasons why this happens and to find other such examples. However, rest assured that this is not generalizable, as we can see from the following example: $(8 + 2)^2 = 10^2 = 100$. This answer is not obtainable by dropping the plus sign and the exponent as before, since $(8 + 2)^2 \neq 82$. You may wish to search for other examples where this weird scheme does work. However, although these examples may be entertaining, they should also be a sign of caution.

We can find other such nifty calculation "shortcuts" that are nice to observe but wrong to generalize. In one case, we have a sum of digits all raised to the same power. Here we just drop the exponents and the plus signs to get the correct answer:

$$1^3 + 5^3 + 3^3 = 153\ (= 1 + 125 + 27).$$

Figure 2-19 shows this principle applied to sums of digits raised to various powers and achieving the "correct" result.

(Sum of the Digits)^n		Number
$1^3 + 5^3 + 3^3$	=	153
$1^1 + 7^2 + 5^3$	=	175
$1^1 + 3^2 + 0^3 + 6^4$	=	1306
$8^4 + 2^4 + 0^4 + 8^4$	=	8,208
$4^5 + 1^5 + 5^5 + 1^5$	=	4,151
$3^3 + 4^4 + 3^3 + 5^5$	=	3,435
$2^1 + 6^2 + 4^3 + 6^4 + 7^5 + 9^6 + 8^7$	=	2,646,798

Figure 2-19

Above all, bear in mind that extending this *false rule* would be an embarrassing mistake, as we can see from the following: $1^1 + 2^2 + 3^3 = 1 + 4 + 27 = 32 \neq 123$.

We must always be careful when we discover an amazing relationship, as it may be unique, and not necessarily generalizable. Remember that, by definition, an exception can never be generalized, or else a colossal mistake will result.

Try to Find the Mistake Here—A Paradox

A customer walks into a bookshop and buys a book for \$10. The next day, he returns to the bookshop and returns the book he bought the previous day. He then selects a book costing \$20 and walks out with it. He reasons that he paid for the \$10 book on the first day, and then returned the \$10 book, thereby giving the bookshop \$10 cash plus the \$10 book. With this \$20 credit, he took a \$20 book and considered it an even trade. Is this correct? If not, where is the error? There is obviously a subtle mistake made for the reader to discover. (*Hint*: Try doing this by replacing "a \$10 book" with "two \$5 bills." The mistake should then become clear.)

The Paradox of the Missing Dollar

Three men plan to spend one night in a hotel room. They pay $60 for the hotel room. Just as they are about to check out, the receptionist notices that the cost for the room was $55 per night. The receptionist sends the bellhop to the room to return the $5 of overpayment. However, the bellhop decides to give each of the three guests $1, and keeps the remaining $2 for himself. Therefore, each of the three guests has paid only $19 for the room. The sum of these three payments is therefore 3 × $19 = $57. This plus the $2 that the bellhop kept only totals to $59. Where is the missing dollar? Is there some mistake?

After a somewhat bewildered reaction to this transaction, we offer the following explanation of the mistake: It is totally meaningless to add the $2 taken by the bellhop to the $57 paid by the three men. The correct calculation is as follows: Three men paid $57 for the room, of which $55 went to the receptionist and $2 went to the bellhop.

Another way of looking at this (unmistakenly) is to note that the three guests got a refund of $3. Add to this the $55 price for the room and the $2 the bellhop took to find a total of $60. Such calculating mistakes are not uncommon, but should not be accepted casually.

A Mistake in Averaging Rates

A common mathematics mistake occurs when we seek the average speed over a round trip. For example, let's assume that the speed going to a destination is calculated at 60 mph, and the speed of the return trip over the same route is 30 mph. The typical mistake here is to take the two speeds and treat them with equal weight. That is, to add them and then divide by 2 to get an average round-trip speed of 45 mph. Although this would be the correct procedure for taking the arithmetic mean (the common average) of two quantities of *equal* value, it is not the correct way to treat two quantities of *unequal* value—in this case, the rates taken over two *different* amounts of time.

This is the case for the two speeds being considered, since it takes twice as long to travel a distance at 30 mph than it does to travel the same distance at 60 mph. Consequently, the 30 mph speed must be given double the weight. Therefore, the correct average speed for the entire roundtrip would be calculated as follows:

$$\frac{60+30+30}{3}=40 \text{ mph.}$$

To calculate the average of speeds that are less convenient than the two we used in the above example, we would employ the concept of a *harmonic mean*. The harmonic mean is defined as "the reciprocal of the average of the reciprocals" of the values being considered. That is, for two values a and b, the harmonic mean is

$$\frac{1}{\frac{\frac{1}{a}+\frac{1}{b}}{2}}=\frac{2}{\frac{1}{a}+\frac{1}{b}}=\frac{2ab}{a+b}.$$

For the above example, we can use the formula for the harmonic mean to get the average speed as follows:

$$\frac{2\cdot 60\cdot 30}{60+30}=40.$$

The harmonic mean becomes particularly useful when the numbers to be averaged are more inconvenient. Suppose we now use as our roundtrip speeds 58 mph and 32 mph. Applying the harmonic mean formula to these two speeds, we get

$$\frac{2\cdot 58\cdot 32}{58+32}=41.2\overline{4}.$$

Using the harmonic mean formula also allows us to find the average speed of more than two speeds. When we apply the definition of the harmonic mean, namely, the reciprocal of the average of the reciprocals, to three items, we get the following:

$$\frac{1}{\frac{\frac{1}{a}+\frac{1}{b}+\frac{1}{c}}{3}}=\frac{3}{\frac{1}{a}+\frac{1}{b}+\frac{1}{c}}=\frac{3abc}{bc+ac+ab}.$$

And for four items:

$$\frac{1}{\frac{\frac{1}{a}+\frac{1}{b}+\frac{1}{c}+\frac{1}{d}}{4}}=\frac{4}{\frac{1}{a}+\frac{1}{b}+\frac{1}{c}+\frac{1}{d}}=\frac{4abcd}{bcd+acd+abd+abc}.$$

This can, of course, be expanded when five or more speeds over an equal distance need to be averaged. With this relationship, we can avoid one of the most common mistakes in mathematical calculations.

It should also be noted that the average speed can also be calculated by finding the total distance and dividing it by the total time for the entire trip. Mistakes in arithmetic are not only annoying but also ruin some fine mathematics. Therefore, it is important to bear in mind the many arithmetic mistakes we have highlighted in this chapter.

An Arithmetic Challenge

Applying reasoning skills to an addition algorithm situation can be very important in understanding arithmetic processes and training your mathematical thinking. Consider the following problem.

The following letters represent the digits of a simple addition:

$$\begin{array}{r} \mathbf{S\ E\ N\ D} \\ \underline{+\ \ \mathbf{M\ O\ R\ E}} \\ \mathbf{M\ O\ N\ E\ Y} \end{array}$$

Find the digits that represent the letters to make this addition correct. Show that the solution is unique. Most important in this activity is the analysis, and particular attention should be given to the reasoning you use. At first, this may appear to be a daunting task, but if you take it step by step, you will find it rewarding and entertaining.

The sum of two four-digit numbers cannot yield a number greater than 19,999. Therefore, **M** = **1**.

We then have MORE < 2000 and SEND $< 10{,}000$. It follows that MONEY $< 12{,}000$. Thus, O can be either 0 or 1. But the 1 is already used; therefore, **O = 0.**

We now have:

$$\begin{array}{r} \mathbf{S\ E\ N\ D} \\ .\quad \mathbf{1\ 0\ R\ E} \\ \hline \mathbf{1\ 0\ N\ E\ Y} \end{array}$$

Now, MORE < 1100. If SEND is less than 9000, then MONEY $< 10{,}100$, which would imply that $N = 0$. But this cannot be, since 0 was already used; therefore, SEND > 9000, so that **S = 9**. We now have:

$$\begin{array}{r} \mathbf{9\ E\ N\ D} \\ .\quad \mathbf{1\ 0\ R\ E} \\ \hline \mathbf{1\ 0\ N\ E\ Y} \end{array}$$

The remaining digits from which we may complete the problem are {2,3,4,5,6,7,8}.

Let us examine the units digits. The greatest sum of our remaining digits is $7 + 8 - 15$, and the least sum is $2 + 3 = 5$.

If $D + E < 10$, then $D + E = Y$ with no carry over into the tens column. Otherwise, $D + E = Y + 10$, with a 1 carried over to the tens column.

Taking this argument one step further to the tens column, we get $N + R = E$, with no carryover, or $N + R = E + 10$, with a carryover of 1 to the hundreds column. However, if there is no carryover to the hundreds column, then $E + 0 = N$, which implies that $E = N$. This is not permissible. Therefore, there must be a carryover to the hundreds column, so $N + R = E + 10$, and $E + 0 + 1 = N$, or $E + 1 = N$.

We substitute this value for N into the previous equation to get $(E + 1) + R = E + 10$, which implies that $R = 9$. But 9 has already been used for the value of S. We must try a different approach.

We shall therefore assume that $D + E = Y + 10$, since we apparently need a carryover into the tens column, where we just reached a dead end.

Now the sum in the tens column is $1 + 2 + 3 < 1 + N + R < 1 + 7 + 8$. If, however, $1 + N + R < 10$, there will be no carryover to the hundreds column, leaving the previous dilemma of $E = N$, which is not allowed. We then have $1 + N + R = E + 10$, which insures the needed carryover to the hundreds column.

There, $1 + E + 0 = N$, or $E + 1 = N$. Substituting this in the above equation $(1 + N + R = E + 10)$ gives us $1 + (E + 1) + R = E + 10$, or **R = 8**.

We now have:

$$\begin{array}{r} 9\ E\ N\ D \\ .\quad 1\ 0\ 8\ E \\ \hline 1\ 0\ N\ E\ Y \end{array}$$

From the remaining list of available digits, we find that $D + E < 14$. From the equation $D + E = Y + 10$, Y is either 2 or 3. If $Y = 3$, then $D + E = 13$, implying that the digits D and E can take on only 6 or 7. If $D = 6$ and $E = 7$, then from the previous equation $E + 1 = N$, we would have $N = 8$, which is unacceptable since $R = 8$. If $D = 7$ and $E = 8$, then from the previous equation $E + 1 = N$, we would have $N = 9$, which is unacceptable since $S = 9$. Therefore, **Y = 2**.

We now have:

$$\begin{array}{r} 9\ E\ N\ D \\ .\quad 1\ 0\ 8\ E \\ \hline 1\ 0\ N\ E\ 2 \end{array}$$

Thus, $D + E = 12$. The only way to get this sum is with 5 and 7. If $E = 7$, we once again get from $E + 1 = N$ the contradictory $N = 8$, which is not acceptable. Therefore, **D = 7** and **E = 5**. We can now again use the equation $E + 1 = N$ to get **N = 6**.

Finally, we get the solution:

$$\begin{array}{r} 9\ 5\ 6\ 7 \\ .\quad 1\ 0\ 8\ 5 \\ \hline 1\ 0\ 6\ 5\ 2 \end{array}$$

This rather strenuous activity should strengthen one's facility in arithmetic.

Sum of Unit Fractions

In ancient Egypt, aside from the fraction $\frac{2}{3}$, all measurements were made by using unit fractions. Therefore, it was important to determine how fractional measurements can be expressed as the sum of various unit fractions. Our challenge here is to show that all unit fractions between 0 and 1 of the form $\frac{1}{n}$ $(n \geq 2)$ and fractions of the form $\frac{2}{n}$ $(n \geq 3)$ can be represented as sums of different unit fractions. (Note: the representation

$$\frac{2}{n} = \frac{1}{n} + \frac{1}{n}$$

is not acceptable because the unit fractions are not different, as required.)

Solution: One can easily check some small numbers for n:

$$\frac{1}{2} = \frac{1}{3} + \frac{1}{6}, \frac{1}{3} = \frac{1}{4} + \frac{1}{12}, \frac{1}{4} = \frac{1}{12} = \frac{1}{5} + \frac{1}{20} \cdots$$

Sometimes, a fraction can be represented as more than one such sum of different unit fractions, as we have shown above with the fraction $\frac{1}{4}$. But we are challenged to prove that there is always such a representation available. Suppose we start in the case of $\frac{1}{n}$ with the largest possible unit fraction as the first summand:

$$\frac{1}{n} = \frac{1}{n+1} + x.$$

And we are lucky, as x will turn out as a *different* unit fraction:

$$x = \frac{1}{n} - \frac{1}{n+1} = \frac{n+1-n}{n(n+1)} = \frac{1}{n(n+1)}.$$

(Note: for $n \geq 2$ we have $\frac{1}{n(n+1)} < \frac{1}{n+1}$.) Thus, we have demonstrated what was requested.

For $\frac{2}{n}$ $(n \geq 3)$, we can distinguish two cases. If, on the one hand, n is an even number, say, $n = 2k$, then $\frac{2}{n} = \frac{1}{k}$. If, on the other hand, n is an odd number, say, $n = 2k - 1$, then

$$\frac{2}{n} = \frac{2}{2k-1}.$$

When we use the same technique as above and try the biggest possible unit fraction, which here is $\frac{1}{k}$, as the first summand, we are successful, as the remaining summand must then be a unit fraction, too:

$$\frac{2}{2k-1} - \frac{1}{k} = \frac{1}{k(2k-1)}$$

and thus

$$\frac{2}{2k-1} = \frac{1}{k} + \frac{1}{k(2k-1)}.$$

Arithmetic Peculiarities

As we complete the chapter on arithmetic procedures and peculiarities, let's consider some of the following peculiarities, which tend to speak for themselves (Figures 2-20–2-27):

987654321	×	9	=	8 888 888 889
987654321	×	18	=	17 777 777 778
987654321	×	27	=	26 666 666 667
987654321	×	36	=	35 555 555 556
987654321	×	45	=	44 444 444 445
987654321	×	54	=	53 333 333 334
987654321	×	63	=	62 222 222 223
987654321	×	72	=	71 111 111 112
987654321	×	81	=	80 000 000 001

Figure 2-20

12345679	×	9	=	111,111,111
12345679	×	18	=	222,222,222
12345679	×	27	=	333,333,333
12345679	×	36	=	444,444,444
12345679	×	45	=	555,555,555
12345679	×	54	=	666,666,666
12345679	×	63	=	777,777,777
12345679		72	=	888,888,888
12345679		81	=	999,999,999

Figure 2-21

$$
\begin{array}{rcccccl}
0 & \times & 9 & + & 1 & = & 1 \\
1 & \times & 9 & + & 2 & = & 11 \\
12 & \times & 9 & + & 3 & = & 111 \\
123 & \times & 9 & + & 4 & = & 1{,}111 \\
1{,}234 & \times & 9 & + & 5 & = & 11{,}111 \\
12{,}345 & \times & 9 & + & 6 & = & 111{,}111 \\
123{,}456 & \times & 9 & + & 7 & = & 1{,}111{,}111 \\
1{,}234{,}567 & \times & 9 & + & 8 & = & 11{,}111{,}111 \\
12{,}345{,}678 & \times & 9 & + & 9 & = & 111{,}111{,}111
\end{array}
$$

Figure 2-22

$$
\begin{array}{rcccl}
999{,}999 & \times & 1 & = & 0{,}999{,}999 \\
999{,}999 & \times & 2 & = & 1{,}999{,}998 \\
999{,}999 & \times & 3 & = & 2{,}999{,}997 \\
999{,}999 & \times & 4 & = & 3{,}999{,}996 \\
999{,}999 & \times & 5 & = & 4{,}999{,}995 \\
999{,}999 & \times & 6 & = & 5{,}999{,}994 \\
999{,}999 & \times & 7 & = & 6{,}999{,}993 \\
999{,}999 & \times & 8 & = & 7{,}999{,}992 \\
999{,}999 & \times & 9 & = & 8{,}999{,}991 \\
999{,}999 & \times & 10 & = & 9{,}999{,}990
\end{array}
$$

Figure 2-23

$$
\begin{array}{rcccccl}
 & \times & 9 & + & 8 & = & 8 \\
9 & \times & 9 & + & 7 & = & 88 \\
98 & \times & 9 & + & 6 & = & 888 \\
987 & \times & 9 & + & 5 & = & 8{,}888 \\
9{,}876 & \times & 9 & + & 4 & = & 88{,}888 \\
98{,}765 & \times & 9 & + & 3 & = & 888{,}888 \\
987{,}654 & \times & 9 & + & 2 & = & 8{,}888{,}888 \\
9{,}876{,}543 & \times & 9 & + & 1 & = & 88{,}888{,}888 \\
98{,}765{,}432 & \times & 9 & + & 0 & = & 888{,}888{,}888
\end{array}
$$

Figure 2-24

$$
\begin{array}{rclcl}
9 & \times & 9 & = & 81 \\
99 & \times & 99 & = & 9{,}801 \\
999 & \times & 999 & = & 998{,}001 \\
9{,}999 & \times & 9{,}999 & = & 99{,}980{,}001 \\
99{,}999 & \times & 99{,}999 & = & 9{,}999{,}800{,}001 \\
999{,}999 & \times & 999{,}999 & = & 999{,}998{,}000{,}001 \\
9{,}999{,}999 & \times & 9{,}999{,}999 & = & 99{,}999{,}980{,}000{,}001
\end{array}
$$

Figure 2-25

$$37037 \times 3 = 111111$$
$$37037 \times 6 = 222222$$
$$37037 \times 9 = 333333$$
$$37037 \times 18 = 666666$$
$$37037 \times 12 = 444444$$
$$37037 \times 15 = 555555$$
$$37037 \times 18 = 666666$$
$$37037 \times 21 = 777777$$
$$37037 \times 24 = 888888$$
$$37037 \times 27 = 999999$$
$$37037 \times 30 = 1111110$$
$$37037 \times 33 = 1222221$$
$$37037 \times 36 = 1333332$$
$$37037 \times 39 = 1444443$$
$$37037 \times 42 = 1555554$$
$$37037 \times 45 = 1666665$$
$$37037 \times 48 = 1777776$$
$$37037 \times 51 = 1888887$$
$$37037 \times 54 = 1999998$$

Figure 2-26

1	=	1	=	1^2
1+2+1	=	2+2	=	2^2
1+2+3+2+1	=	3+3+3	=	3^2
1+2+3+4+3+2+1	=	4+4+4+4	=	4^2
1+2+3+4+5+4+3+2+1	=	5+5+5+5+5	=	5^2
1+2+3+4+5+6+5+4+3+2+1	=	6+6+6+6+6+6	=	6^2
1+2+3+4+5+6+7+6+5+4+3+2+1	=	7+7+7+7+7+7+7	=	7^2
1+2+3+4+5+6+7+8+7+6+5+4+3+2+1	=	8+8+8+8+8+8+8+8	=	8^2
1+2+3+4+5+6+7+8+9+8+7+6+5+4+3+2+1	=	9+9+9+9+9+9+9+9+9	=	9^2

Figure 2-27

As a last arithmetic curiosity, shown in Figure 2-28, we need to find an eight-digit number in which no digit is repeated, and which, when multiplied by 9, yields a nine-digit number in which no digit is repeated. Here are a few such possibilities:

$$81{,}274{,}365 \times 9 = 731{,}469{,}285$$
$$72{,}645{,}831 \times 9 = 653{,}812{,}479$$
$$58{,}132{,}764 \times 9 = 523{,}194{,}876$$
$$76{,}125{,}483 \times 9 = 685{,}129{,}347$$

Figure 2-28

Chapter 3

The Spectacular of Geometry

The visual aspects of mathematics are centered on geometry. Unfortunately, when geometry is taught in schools, the focus is not on showing the beauty of the subject but rather its usefulness in all fields of mathematics. In the United States, high school students spend one year studying geometry and proving various geometric relationships. This distracts from the visual beauty of the subject, as students concentrate on the proof rather than on what has been proved. This chapter is intended to show aspects of geometry that are truly amazing and glorify the beauty of mathematics. In order not to distract from the visual appreciation of these incredible geometric relationships, proofs for these amazing geometric relationships are in the Appendix.

The Varignon Parallelogram

In the world of quadrilaterals, the parallelogram dominates with its most unusual properties. Once again, the most amazing characteristic of parallelograms is typically not taught to high school students. We can create a parallelogram when we join consecutively the midpoints of *any randomly drawn* quadrilateral, as we can see in Figure 3-1, where the midpoints *K*, *L*, *M*, and *N* of the sides *AB*, *BC*, *CD*, and *AD* are joined consecutively to form parallelogram *KLMN*. This parallelogram is often referred to as a *Varignon parallelogram*, named after the French mathematician Pierre Varignon (1654–1722).

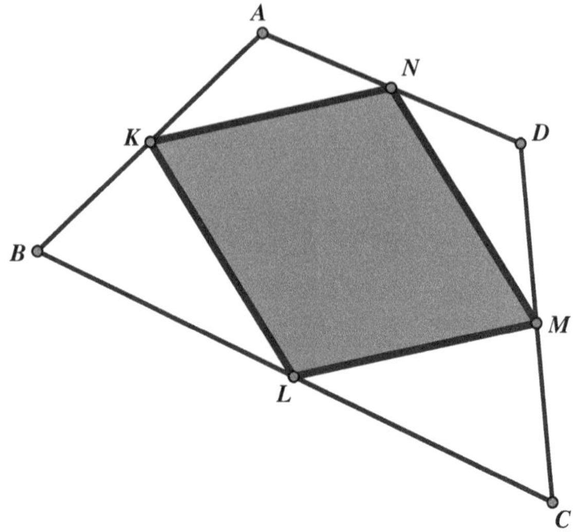

Figure 3-1

The Varignon Parallelogram's Special Properties

The Varignon parallelogram has other special properties, which we illustrate in Figure 3-2. For example, the perimeter of a Varignon

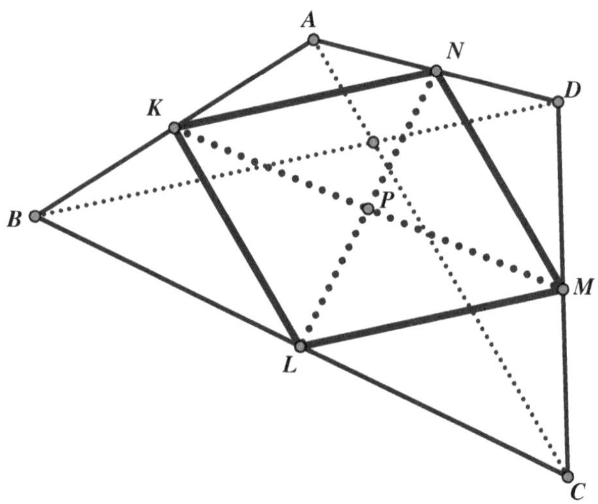

Figure 3-2

parallelogram is equal to the sum of the diagonals of the original quadrilateral. Furthermore, the Varignon parallelogram is half the area of the original quadrilateral. Another feature of a quadrilateral is that the lines joining the midpoints of the opposite sides bisect each other as they are actually also the diagonals of the Varignon parallelogram. This is just a beginner's look at that very special parallelogram. An ambitious reader may search for other surprising relationships.

An Alternative Way of Creating a Varignon Parallelogram

To conclude our little feature on the special Varignon parallelogram, follow along as we consider a rather strange—yet simple—way to create such a parallelogram. In Figure 3-3, two unrelated triangles *ABC* and *DBC* that share the same base *BC* can also determine a parallelogram *PQRS* by joining the midpoints of the sides *P*, *Q*, *R*, and *S*.

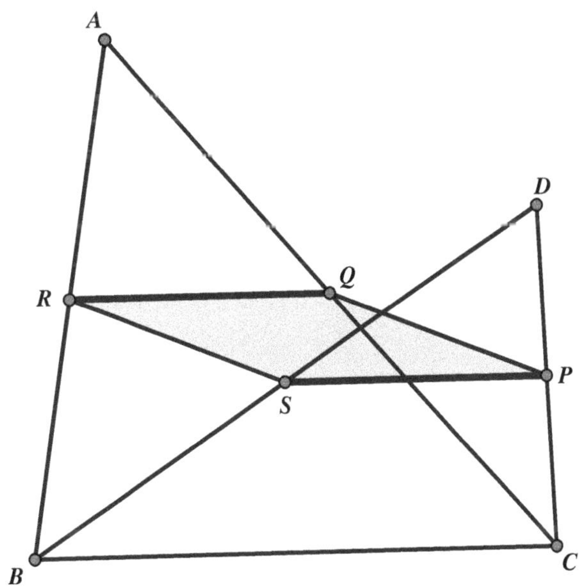

Figure 3-3

A Random Point in a Parallelogram Determines Equal Area Triangle Relationships

Any point in a parallelogram has a special feature that is very rarely popularized but shows unexpected relationships. Consider parallelogram *ABCD* in Figure 3-4, where point *P* is simply a random point in the interior of the parallelogram. The lines joining *P* to the vertices of the parallelogram determine four triangles: $Area\Delta APB$, $Area\Delta DPC$, $Area\Delta BPC$, and $Area\Delta APD$. Completely unexpectedly, we find that the sums of the areas of the pairs of opposite triangles are equal. That is, $Area\Delta APB + Area\Delta DPC = Area\Delta BPC + Area\Delta APD$.

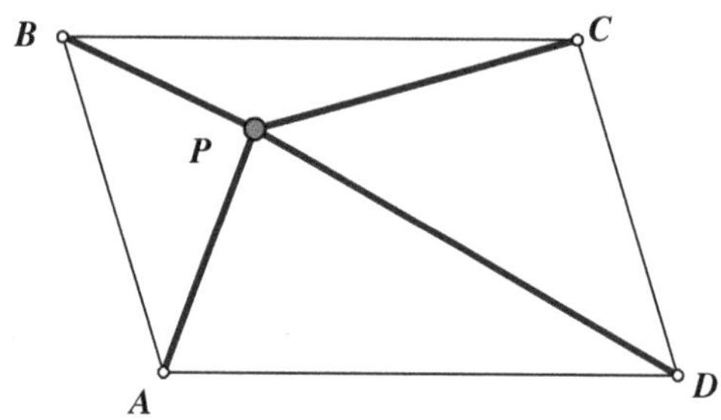

Figure 3-4

Unusual Parallelogram Comparisons

Here is a problem that has perplexed scores of high school teachers, who found it in a 150-year-old geometry book. On the surface, the problem looks very difficult given the procedures that are typically required to solve geometry problems. However, the solution becomes quite simple when one thinks "out-of-the-box." This is what makes the problem attractive to those who dislike learning about parallelograms. We are given parallelograms *ABCD* and *APQR*, with point *P* on side *BC* and point *D* on side *RQ*, as shown in Figure 3-5. If the area of parallelogram *ABCD* is 18, what is the area of parallelogram *APQR*?

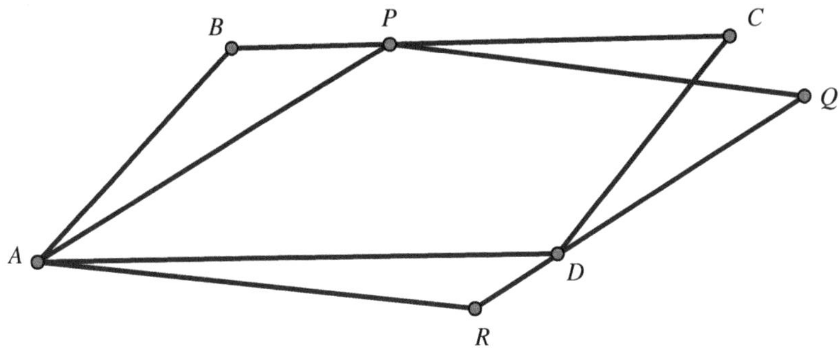

Figure 3-5

The Unexpected Triangle That Is Half the Area of a Trapezoid

Certain special quadrilaterals allow themselves to be partitioned into two equal areas. The simplest one is any type of parallelogram, such as the square, rectangle, or rhombus, where a diagonal creates two equal-area triangles. However, it is not so easy to find a triangle that is half the area of a given trapezoid. The procedure is rather simple. Figure 3-6 shows trapezoid *ABCD*, with parallel sides *AB* and *CD*. All we need to do is locate the midpoint *P* of side *AD*, and then draw *BP* and *CP*. The result is that area $\Delta BCP = \frac{1}{2}$area parallelogram *ABCD*.

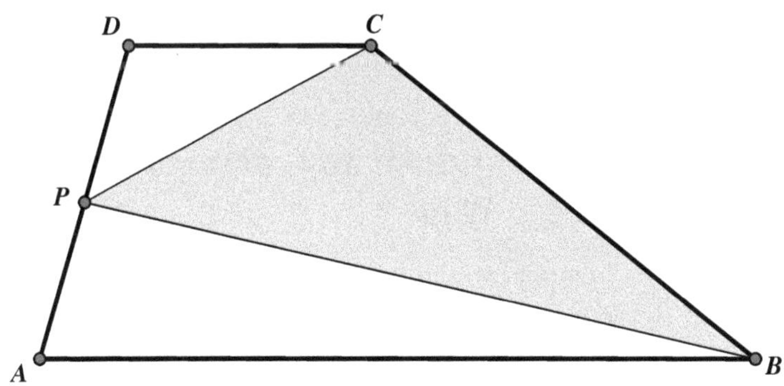

Figure 3-6

A Truly Unexpected Collinearity

Drawing an equilateral triangle on the inside of a given square *ABCD*, and another equilateral triangle on the outside of the square, as shown in Figure 3-7, provides an interesting and unexpected collinearity of three points. The two equal equilateral triangles *BCE* and equilateral triangle *DCF* are placed inside and outside of square *ABCD*, respectively. The surprising result is that the points *A*, *E*, and *F* are collinear.

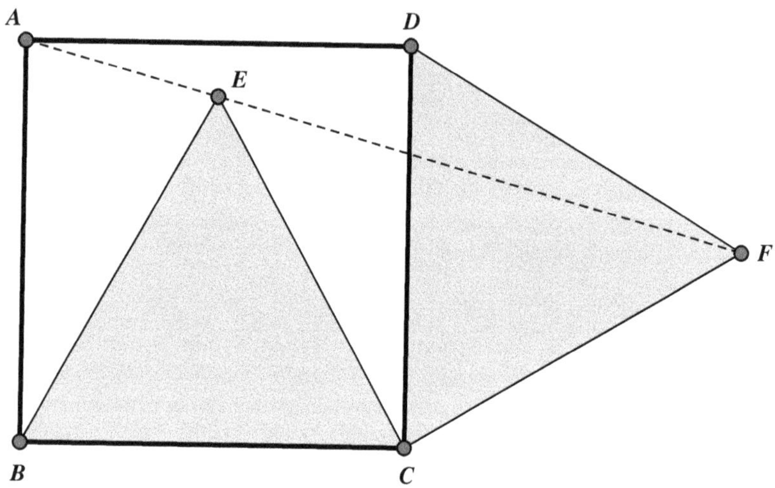

Figure 3-7

Two Equal Circles Produce an Isosceles Triangle in a Rather Strange Way

In Figure 3-8, two equal circles with centers *O* and *Q* intersect at points *A* and *B*. We now draw a line through point *A* that intersects circle *O* at point *X* and intersects circle *Q* at point *Y*. Amazingly, we find that triangle *BYX* is always isosceles, no matter how we draw the line through *A* to intersect the two circles. This is a truly unusual way to construct an isosceles triangle.

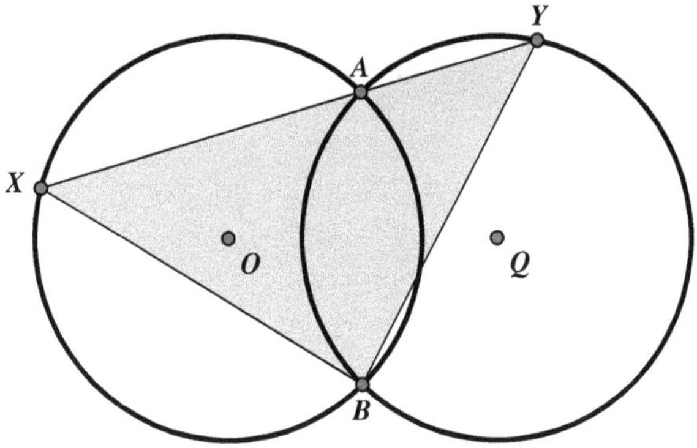

Figure 3-8

Two Equal Circles Generate Another Isosceles Triangle

In Figure 3-9, once again we have two equal circles, with centers *O* and *Q*, intersecting at points *A* and *B*. This time we will form a smaller isosceles triangle in an unexpected fashion. We begin by drawing a line through point *A* that intersects circle *O* at a point *X* (in the interior of circle *Q*) and intersects circle *Q* at a point *Y* (external to circle *O*). Surprisingly, when we draw *BX*, we have created an isosceles triangle *BYX*. This triangle will be isosceles independent of the location of point *X*.

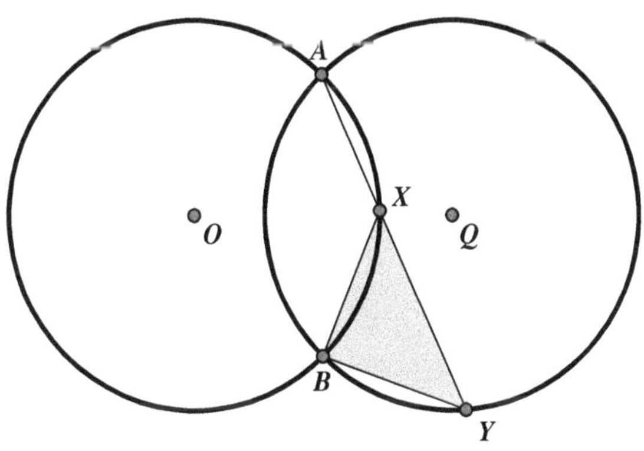

Figure 3-9

Triangles with a Common Base Producing Concurrent Angle Bisectors

Imagine several triangles of different shapes sharing a common base, with the angles opposite the common base equal. Unexpectedly, the bisectors of these equal angles all share a common point. In Figure 3-10, we show four triangles with a common base AB and equal vertex angles at points C, D, E, and F, namely, $\angle ACB = \angle ADB = \angle AEB = \angle AFB$. Quite unexpectedly, we find that the bisectors of each of the equal angles are concurrent at point P.

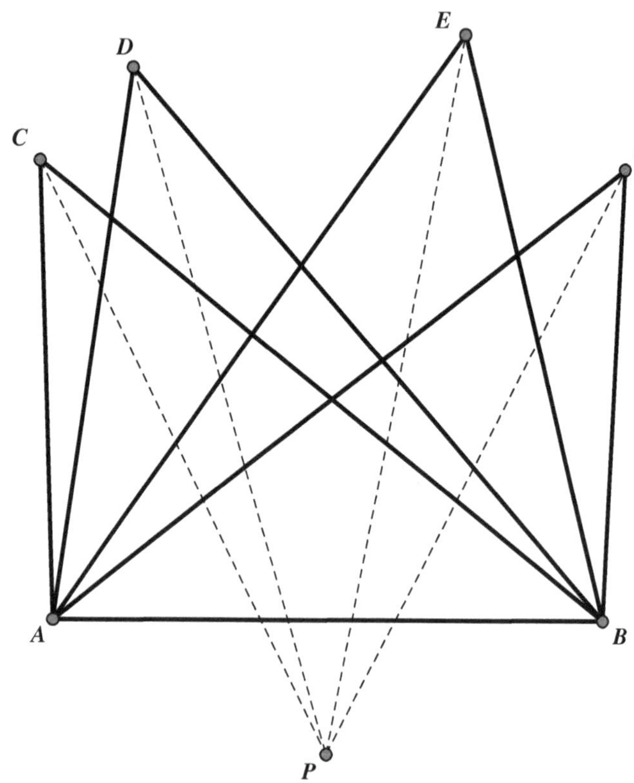

Figure 3-10

Creating Unexpected Supplementary Angles

We offer here a most unusual situation regarding a circle. Suppose you have a circle with two perpendicular chords. We then find that these two chords can generate supplementary central angles of the circle. In Figure 3-11, a circle with center O has chords AB and CD perpendicular at point P. Surprisingly, we find that central angles $\angle AOD$ and $\angle BOC$ are supplementary, that is, $\angle AOD + \angle BOC = 180°$.

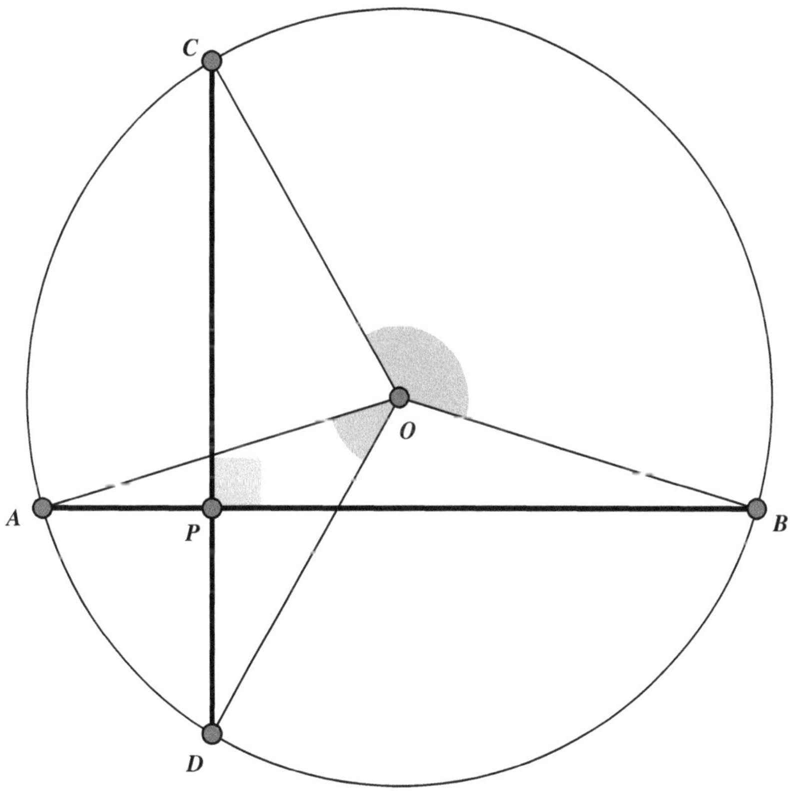

Figure 3-11

The Circumscribable Quadrilateral

A quadrilateral is considered *circumscribable* if it contains a circle that is tangent to all four sides of the quadrilateral. What makes this noteworthy is that the sums of the opposite sides of such a quadrilateral are equal. We can see this in Figure 3-12, where the circle with center O is tangent to the sides AB, BC, CD, and DA of quadrilateral $ABCD$ at points E, F, G, and H, respectively. We then have a circumscribable quadrilateral where the sums of the pairs of opposite sides are equal, that is, $AB + CD = BC + DA$. The converse of this is also true; namely, if the pairs of opposite sides have equal sums, then a circle can be inscribed within the quadrilateral.

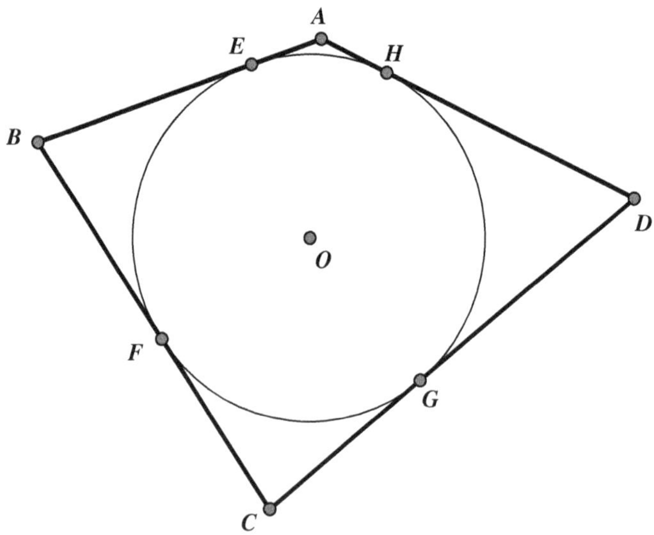

Figure 3-12

Deceptive Geometry Can Be Entertaining

There are times when geometric appearances can be very deceptive. Thinking out-of-the-box is not taught in the high school geometry course, as we will see with the following situation. Squares have been inscribed in congruent isosceles right triangles in two ways, as shown in Figure 3-13. The challenge is to determine which of these squares has a larger area.

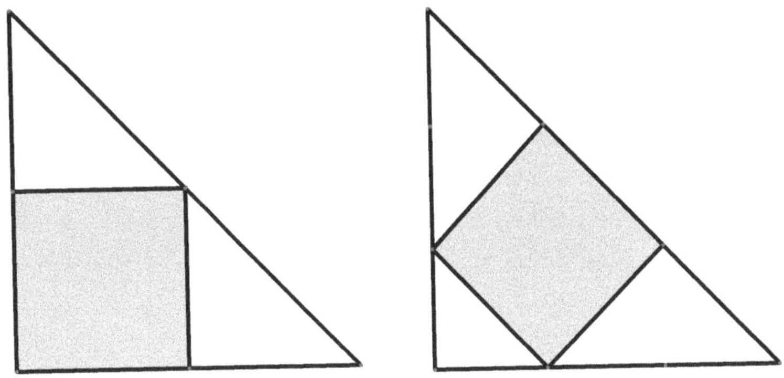

Figure 3-13

The Surprises Embedded in the Reuleaux Triangle

An unusual challenge in geometry is to create a shape that is not a circle but shares the significant properties of a circle. In Figure 3-14, we show a rather odd-looking shape called a *Reuleaux triangle*, named after the German engineer Franz Reuleaux (1829–1905). The Reuleaux triangle is formed by starting with an equilateral

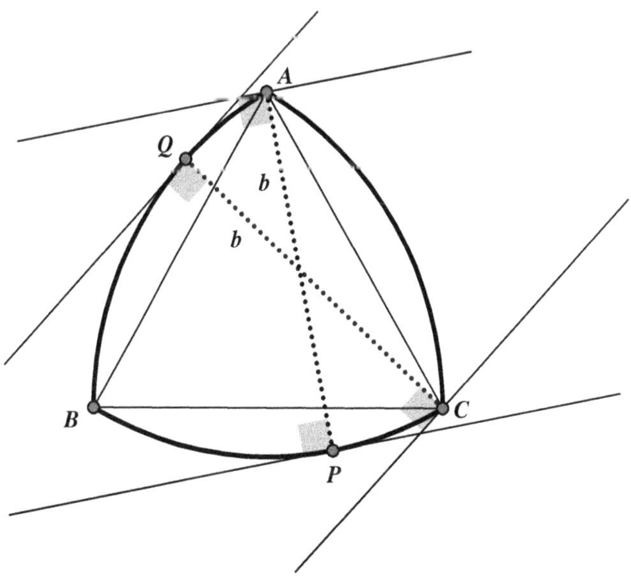

Figure 3-14

triangle ABC and drawing three circular arcs centered at the vertices of ABC with radii equal to the length of the sides of the equilateral triangle.

Now to consider the relationship of this strange looking shape and a circle. Wherever we place a pair of parallel tangents to a circle, they will always be the same distance apart. The same is true for the Reuleaux triangle, where we note this distance b as the breadth of the Reuleaux triangle. In the figure, this constant distance is illustrated by the segments $AP = CQ = b$. If we rotate the Reuleaux triangle, we find that at every instant it will be tangent to the two parallel lines, analogous to rotating a circle between two parallel lines. Surprisingly, the perimeter of the Reuleaux triangle with breadth b is equal to the perimeter of a circle with diameter b.

Semicircles Create the Arbelos

We all know what a semicircle is, but we don't necessarily know the fascinating properties that the semicircle possesses. An example of this is the *arbelos*, which is sometimes referred to as the *shoemaker's knife*. This interesting figure, shown as the lighter shaded region in Figure 3-15, is formed by three tangent semicircles with their

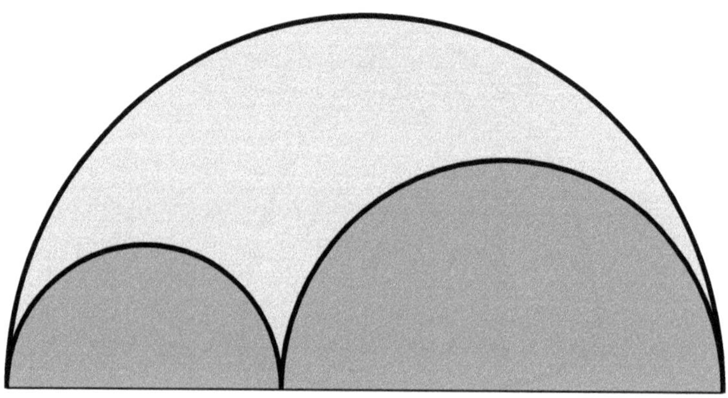

Figure 3-15

diameters on a common line. This means that the sum of the diameters of the two smaller semicircles is equal to the diameter of the larger semicircle. Although it may not be intuitively obvious, the sum of the lengths of the two smaller semicircular arcs is equal to the length of the larger semicircular arc.

An Unexpected Arbelos Bisection

The arbelos harbors an unexpected and strange relationship, which can be seen in Figure 3-16, where adding a few lines to the arbelos results in an interesting feature. Points A and B are the endpoints of the large semicircular arc, and point C is the tangent point of the two smaller semicircular arcs. We construct a line perpendicular to the line segment AB at point C that intersects the larger semicircle at point H. We then draw a common tangent to the two smaller semicircles with tangent points F and G. This tangent intersects segment CH at point S. We then find that the two segments CH and FG bisect each other at point S, so that $SF = SG$ and $SC = SH$.

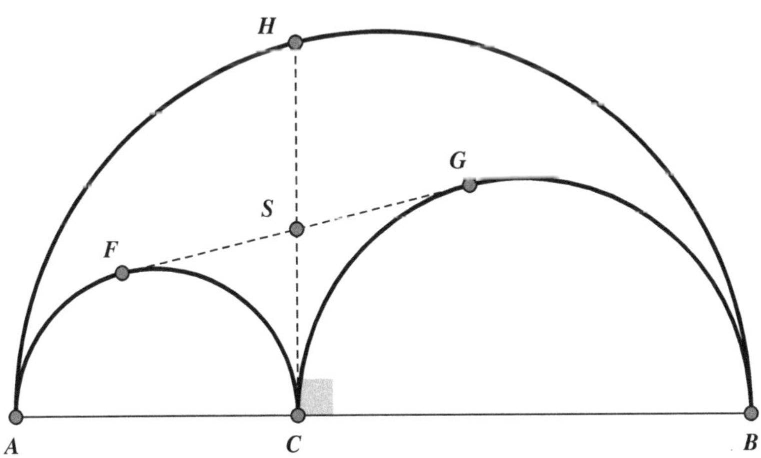

Figure 3-16

More on the Common Arbelos Bisection Point

The arbelos shown in Figure 3-17 yields even more wonderful results; namely, that $HC = FG$, so that point S is the center of a circle containing the points F, C, G, and H. Even more unexpectedly, this circle has the same area as the arbelos, which we recall is the region bounded by the three semicircles.

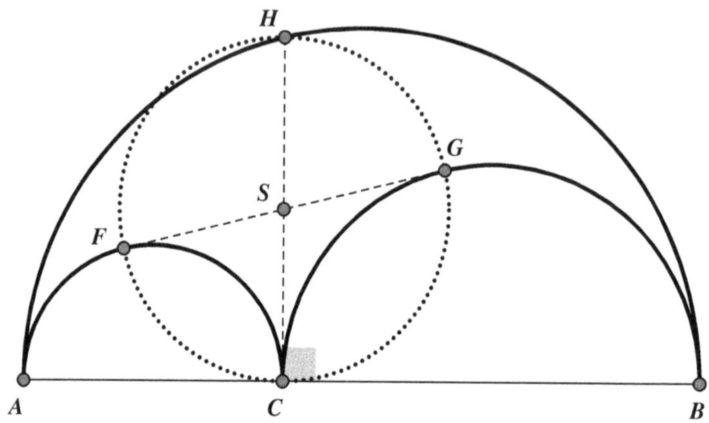

Figure 3-17

The Curious Relationship Between the Orthocenter and Circumcenter

A triangle has a number of important centers. One is the orthocenter, which is the intersection of the triangle's altitudes. Another is the circumcenter, which is the center of circumscribed circle of the triangle. There is a curious relationship involving these two centers that is not well known but easy to appreciate. We show this unexpected relationship in Figure 3-18, where triangle ABC has the orthocenter H and the circumcenter O. Surprisingly, we find that $AH = 2\ OM$, where M is midpoint of AC.

In Figure 3-18, we related the orthocenter to a vertex of the triangle. Now we will switch the relation to have the orthocenter connect to the midpoint M of side BC, an extend the line HM to meet the circumcircle at point P, as shown in Figure 3-19. When we draw the line *CO*, which connects the vertex *C* with the center of the circumcircle, we find that these two lines *HM* and *CO* intersect at point *P* on the circumcircle—a most uncommon concurrency!

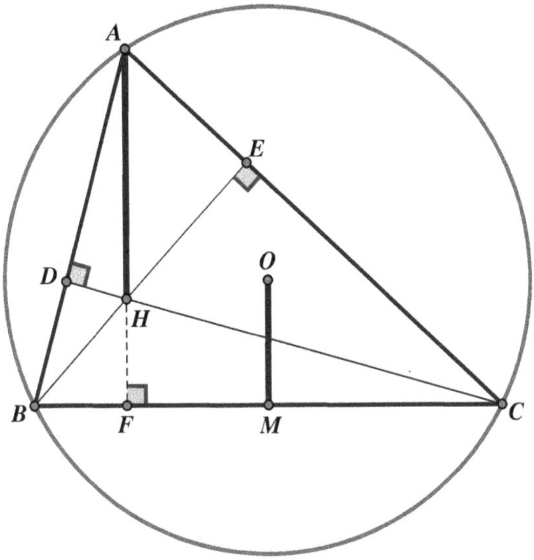

Figure 3-18

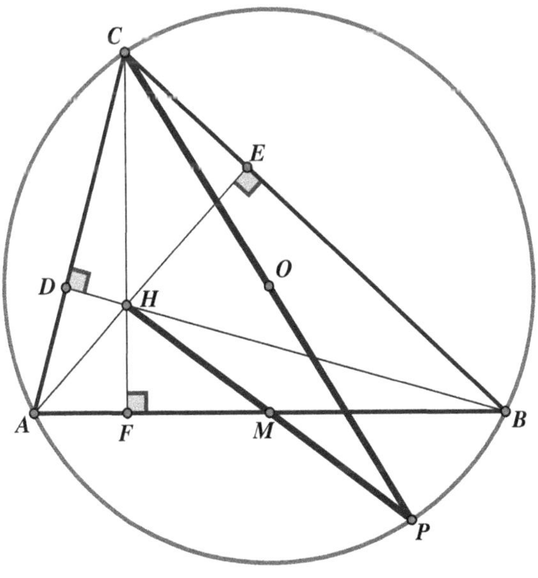

Figure 3-19

The Astonishing Morley's Theorem

Quite often in mathematics, unusual relationships provoke astonishment. In 1900, the mathematician Frank Morley (1860–1937) discovered an amazing relationships that can be applied to any triangle regardless of its shape. Morley demonstrated (and proved) that if we trisect each of the angles of a triangle, the intersections of the adjacent trisectors will always determine an equilateral triangle. Remember, what makes this so amazing is that this relationship holds true regardless of the shape of the original triangle. In Figure 3-20, we have each angle of triangle *ABC* trisected with the adjacent trisectors intersecting at points *D*, *E*, and *F*. When we join these three points, we will always arrive at an equilateral triangle *DEF*.

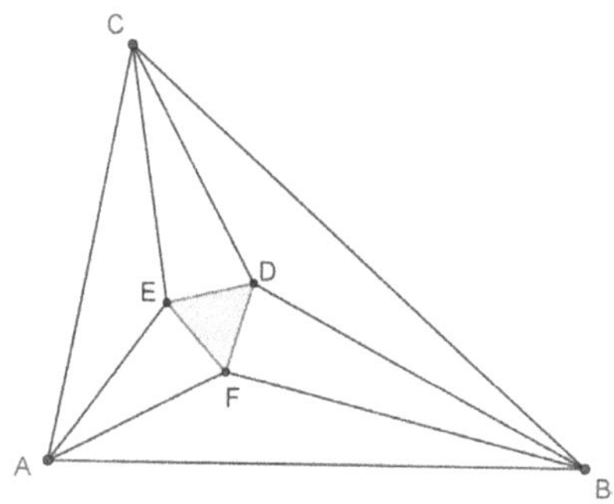

Figure 3-20

To demonstrate the independence of the shape of a triangle that Morley's theorem holds independent of the shape of the triangle, we offer the following variously-shaped triangles in Figures 3-21, 3-22, 3-23, and 3-24.

The astonishing result of this relationship became an international challenge for mathematicians to prove. Consequently, there are now many proofs available, two of which we offer in the Appendix.

Beyond Morley's wonderful discovery, we have found that this configuration also produces another equilateral triangle. As we can see in Figure 3-25, the non-adjacent trisectors of the randomly drawn triangle's angles also intersect to determine an equilateral triangle; that is to say, triangle *PQR* is also equilateral.

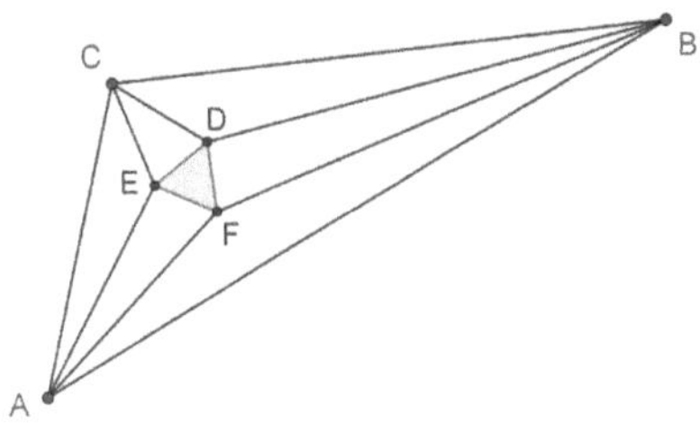

Figure 3-21

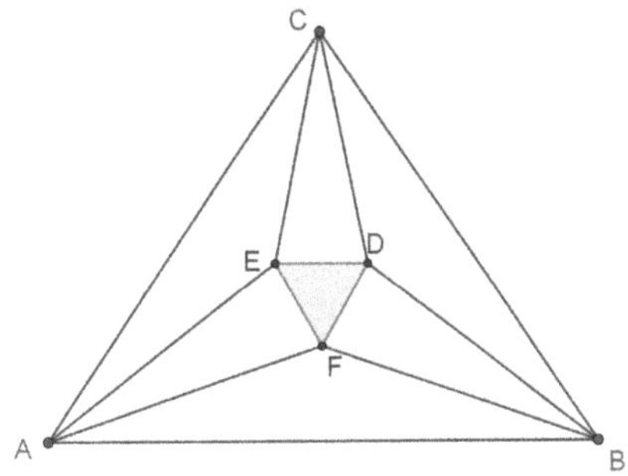

Figure 3-22

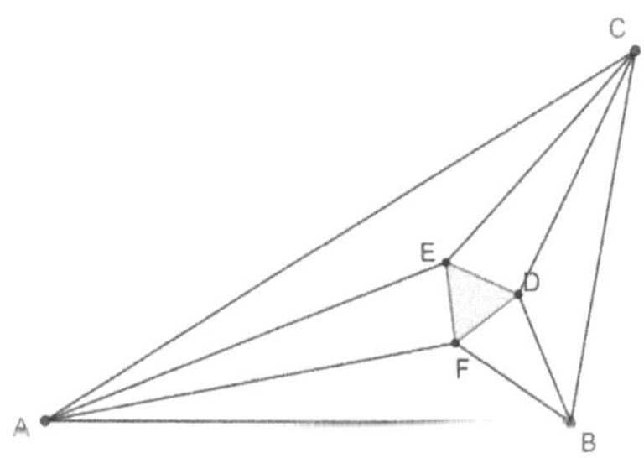

Figure 3-23

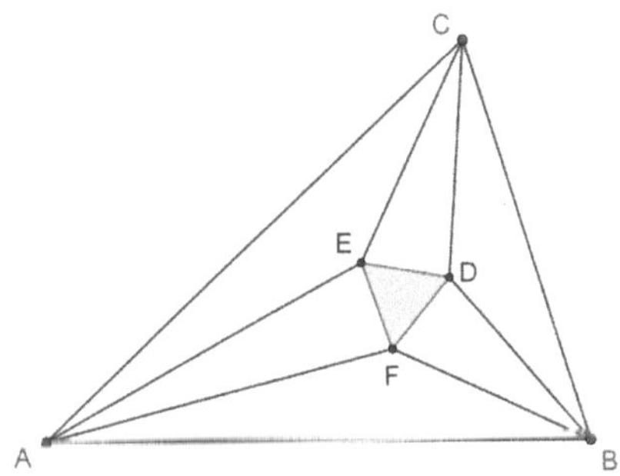

Figure 3-24

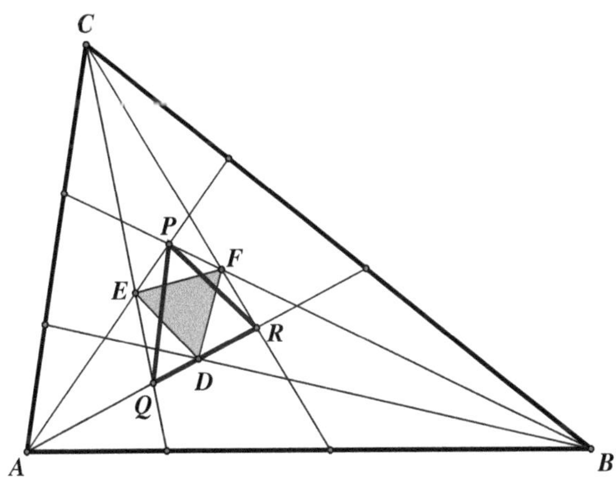

Figure 3-25

Do not be disappointed that we have possibly neglected a concurrency; yes, we have one in this configuration as well. In Figure 3-26, we notice that *CD*, *AF*, and *BE* are concurrent.

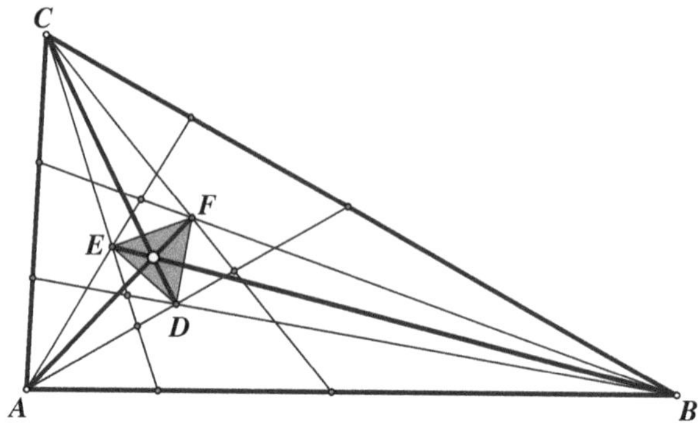

Figure 3-26

Napoleon's Theorem

At this point we are ready to embark on a famous theorem in geometry, one that is attributed to Napoleon Bonaparte (1769–1821). Before becoming famous for his military prowess, Napoleon distinguished himself as a top-flight mathematics student at the Paris Military School and earned membership in the Institut de France, a prestigious scientific society. He prided himself on his talent in mathematics, particularly in geometry. Yet, the theorem that bears his name today was first published by the English mathematician William Rutherford (1798–1871) in "The Ladies' Diary" in 1825, four years after Napoleon's death. To this day, it is quite uncertain if Napoleon ever knew of the relationship we are about to investigate.[1]

[1] See: Grünbaum, B. "Is Napoleon's Theorem *Really* Napoleon's Theorem?" *American Mathematical Monthly* 119, no. 6 (2012), 495–501.

However it was discovered, it is still one of the most surprising relationships in geometry. As shown in Figure 3-27, we begin by drawing a random triangle *ABC*. This triangle can be of any shape. We then construct equilateral triangles on the sides of triangle *ABC*. The triangle formed by joining the three centers of the equilateral triangles (which are the points of intersection of the altitudes, angle bisectors, and medians of the equilateral triangles) is always going to be another equilateral triangle, namely, triangle *KLM*.

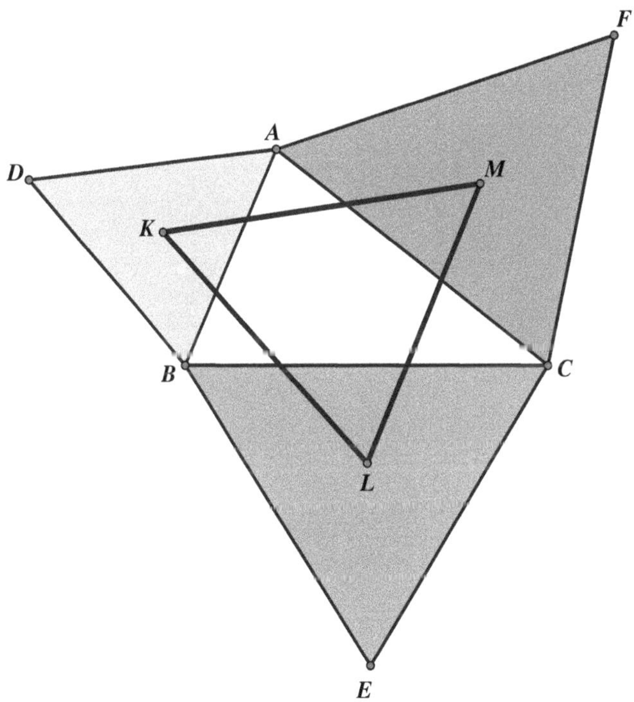

Figure 3-27

The amazement of this arrangement certainly does not end here. When we draw lines from each external vertex of the equilateral triangles to the remote vertex of triangle *ABC*, we find that the three lines are concurrent, as we can see in Figure 3-28, where AE, BF, and CD are concurrent at point P. Furthermore, these three lines are also of equal length, so that $AE = BF = CD$.

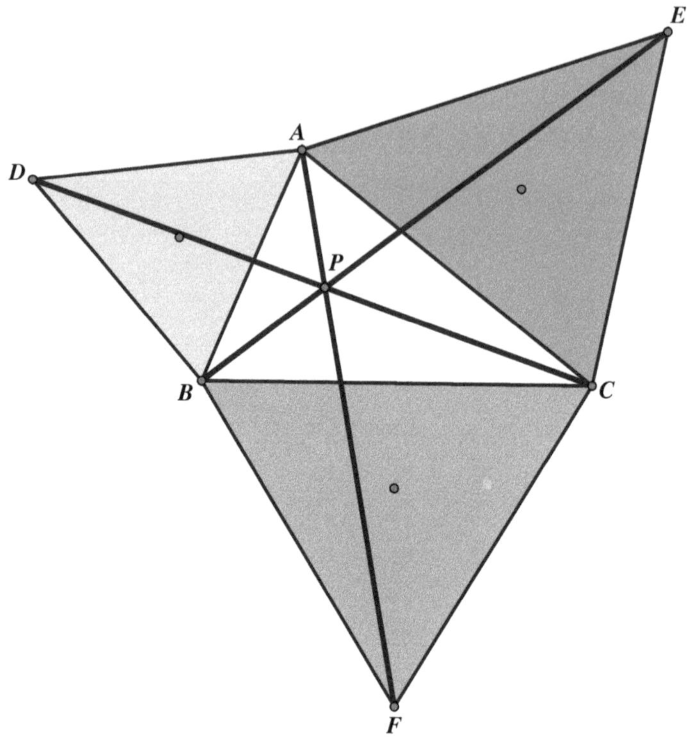

Figure 3-28

If that isn't amazing enough, point P has two surprising properties. First of all, it is the equiangular point of triangle ABC, which means that in Figure 3-29 we have $\angle APB = \angle BPC = \angle CPA = 120°$.

There is much more to be found in this configuration. For example, the point of intersection P, which determined the equiangular point in Figure 3-29, is also the minimum distance point of triangle ABC. That is, the sum of the distances to the vertices from point P, namely, $AP + BP + CP$, is a minimum. Or, we can say that the sum of the

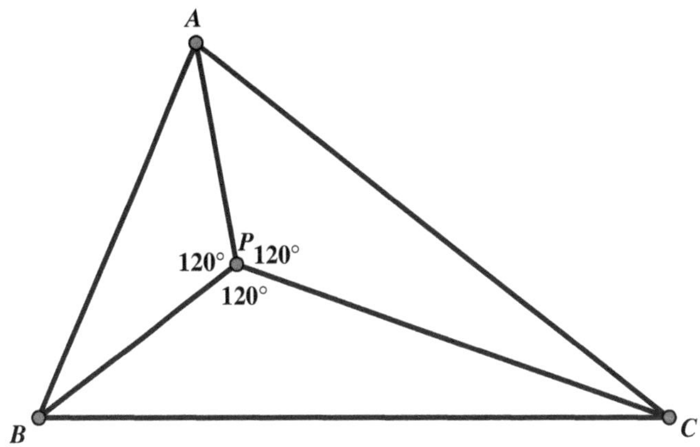

Figure 3-29

distances to the vertices from any other point Q in triangle ABC would be greater than that of point P. That is, in Figure 3-30 we have $QA + QB + QC > PA + PB + PC$.

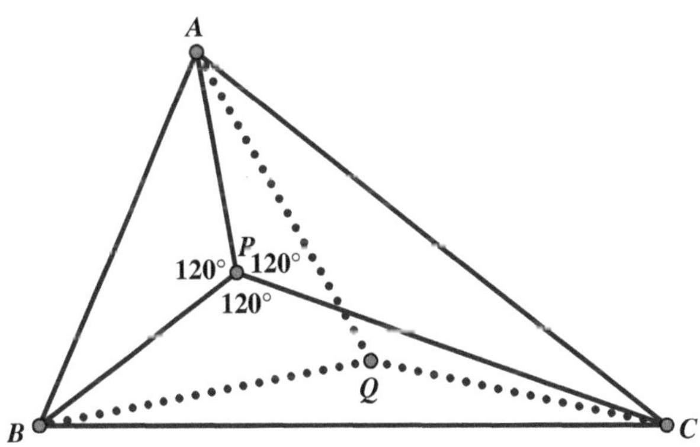

Figure 3-30

There are many more relationships when we construct equilateral triangles on the sides of a random triangle. However, we shall indicate just one more such unusual relationship here. That is, the circumscribed circles of each of these three equilateral triangles are concurrent at the point *P*, at which the lines *AE*, *BF*, and *DC* are also concurrent, as shown in Figure 3-31.

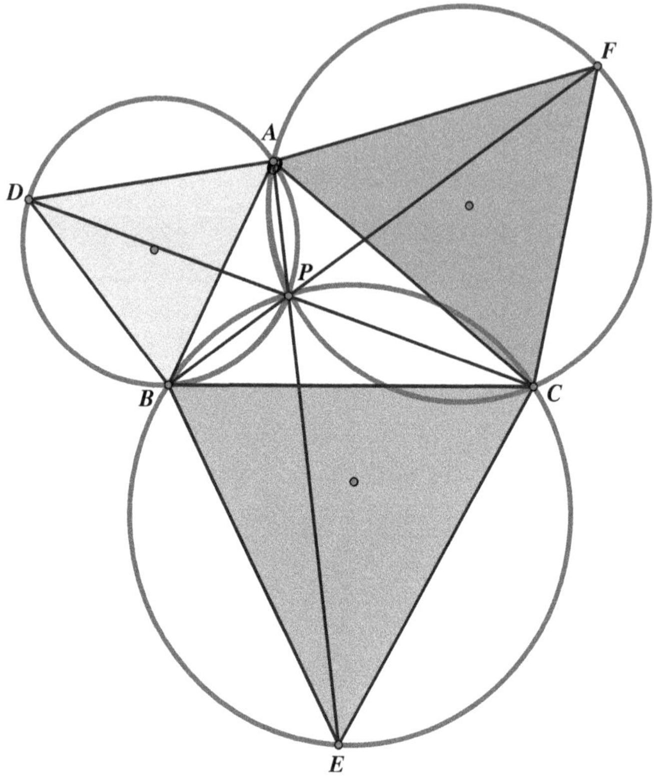

Figure 3-31

The Miquel Point

Circles and triangles together can determine other interesting concurrencies. Consider a randomly selected triangle *ABC* and three randomly selected points, *D*, *E*, and *F*, one on each side of the triangle, as shown in Figure 3-32. We then draw three circles, where each contains two of the three side-points and the included vertex. Lo and behold, the three circles share a common point *M*—the point of concurrency. This point is known as the *Miquel point*, named after the

French mathematician, Auguste Miquel (1816–1851), who published this theorem in *Liouville's Journal* in 1838.

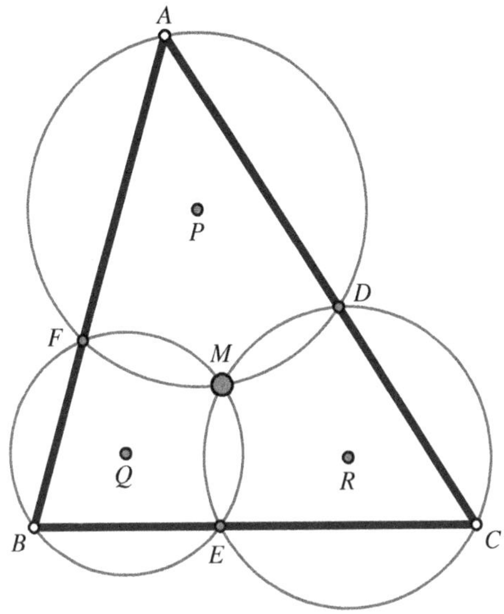

Figure 3-32

Simson's Theorem

When considering collinear points involving triangles, we must acknowledge Simson's theorem. One of the great injustices in the history of mathematics involves this theorem. It was originally published by William Wallace (1768–1843) in Thomas Leybourn's *Mathematical Repository* (1799–1800). Through careless misquotations, the theorem has been attributed to Robert Simson (1687–1768), the famous English interpreter of Euclid's *Elements*, whose book *The Elements of Euclid* formed the basis for the study of geometry in the English-speaking world and greatly influenced the American high school geometry course. To conform to the norm, we shall use the popular reference *Simson's theorem*.

Simson's theorem states that the feet of the perpendiculars drawn from *any* point on the circumscribed circle of a triangle to the sides of the triangle are collinear. This is shown in Figure 3-33, where point *P* is any point on the circumscribed circle of triangle *ABC*. We then draw the perpendiculars from point *P* to each of the sides of triangle *ABC*, so that $PY \perp AC$ at *Y*, $PZ \perp AB$ at *Z*, and $PX \perp BC$ at *X*. According to Simson's theorem, points *X*, *Y*, and *Z* are collinear. This line is usually referred to as the *Simson line*. It should be noted that the converse of this theorem is also true. That is, if the feet of the perpendiculars from a point *P* to the sides of triangle *ABC* are collinear, then point *P* must be on the circumcircle of triangle *ABC*.

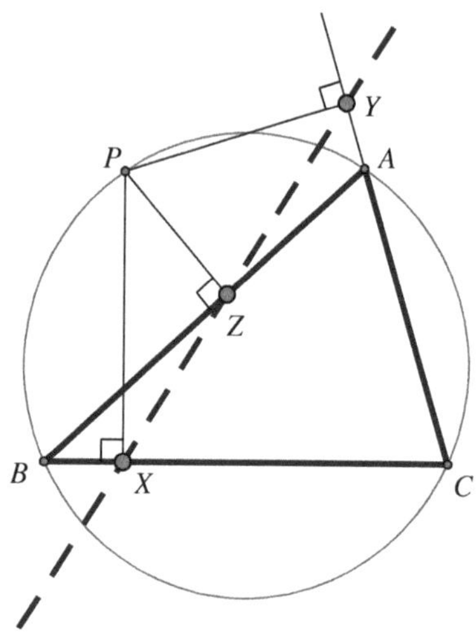

Figure 3-33

The Strange Characteristic of a Hexagon

The number six has a strong presence in a classic theorem named for the French mathematician Blaise Pascal (1623–1662). Consider six points on a circle *c* to be the vertices of a hexagon *ABCDEF* (where the opposite sides are *not* parallel). Amazingly, the three points of intersection of the opposite sides of the hexagon, namely, $P = AB \cap DE$, $Q = BC \cap EF$, and $R = CD \cap FA$, will always lie on a common line *l*. This is shown in Figure 3-34.

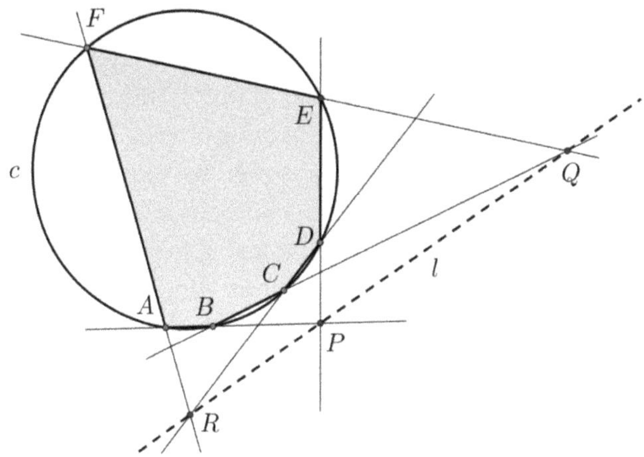

Figure 3-34

The Four-Color Map Problem

We close this chapter with an easily understood geometric phenomenon that was very difficult to justify. It is an "everyday" useful relationship that shows how geometry plays an important role in unanticipated ways. The *Four-Color Map Problem* dates back to 1852, when English mathematician Francis Guthrie (1831–1899), while trying to color a map of the counties of England, noticed that four colors sufficed. He asked his brother Frederick if it was true that *any* map can be colored using only four colors in such a way that adjacent regions, that is, those sharing a common boundary segment and not just a point, receive different colors. Frederick Guthrie then communicated the conjecture to the British mathematician Augustus de Morgan (1806–1871). As early as 1879, the British mathematician Alfred B. Kempe (1849–1922) produced an attempted proof of this conjecture, but in 1890 it was disproven by the British mathematician Percy J. Heawood (1861–1955). Many other subsequent attempts also proved fallacious. Not until 1976 was this so-called "four-color map problem" solved by two mathematicians, Kenneth Appel (1932–2013) and Wolfgang Haken (1928–2022), who, using a computer, considered all possible maps and established that it was never necessary to use more than four colors to color a map so that no two territories sharing a common border would be represented by the same color.[2] They used an IBM 360 that required about 1,200 hours to test the 1,936 cases, which later required inspecting another 1,476 cases. This "computer proof" was not widely accepted by pure mathematicians. However, in 2008, the mathematicians Benjamin Werner and Georges Gonthier produced a formal mathematical solution to the four-color map problem that validated Appel's and Hacken's assertion. The attractive aspect of the four-color map problem lies in the fact that it is easily understood, but the solution has proved to be very elusive and very complex.

[2] Appel, Kenneth; Haken, Wolfgang. "The Solution of the Four-Color-Map Problem." *Scientific American* 237, no. 4 (1977), pp. 108–121.

Chapter 4

Logical Thinking

One of the most important skills that grows with age from childhood on up is logical thinking. Not only is logical thinking an integral part of mathematics, but it is also honed through the study of mathematics. In this chapter, we present problems that can be solved with simple logical thinking, without in-depth knowledge of mathematics. By solving selected problems with just logical thinking, one's ability in this critical area will certainly be improved.

Problem 1: The following problem requires some open thinking and logical reasoning. In the configuration shown in Figure 4-1, there are 11 sticks in each outside row and column. How can you remove one stick from each outside row and one from each outside column, and rearrange the remaining sticks so as to still remain with 11 sticks in each of these rows and columns?

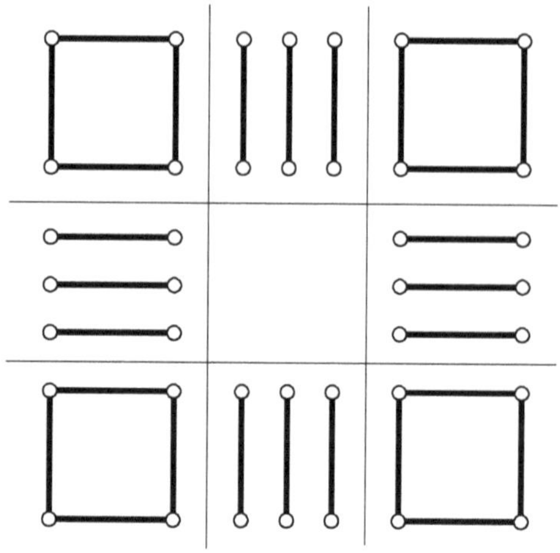

Figure 4-1

Solution: Many people quickly become frustrated as they attempt to solve this problem. They remove one stick from each row and column, and that typically leaves the configuration shown in Figure 4-2.

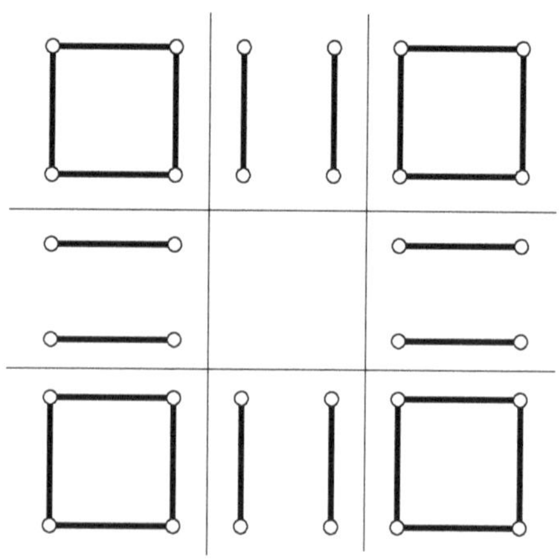

Figure 4-2

However, in Figure 4-2 there remain only 10 sticks in each outside row and outside column. The question is how can we keep 11 sticks in each outside row and outside column after removing one stick from each row and column? Once again, we need to think "out-of-the-box."

We simply take another stick from each row and column, as shown in Figure 4-3, place one at each of the corners, and the problem is solved, as there are now 11 sticks in each of the outside rows and columns. This may appear trivial, but experience shows that most people do not "stumble" easily onto this curious solution. Logical reasoning solves the problem.

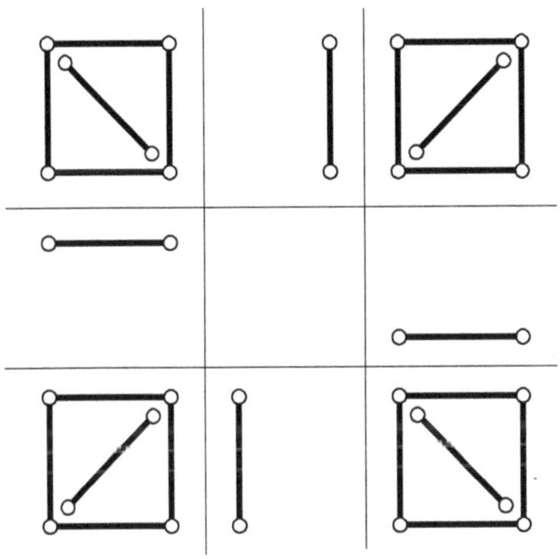

Figure 4-3

Problem 2: Once again logical reasoning is being challenged. We are asked: what is the least number of straight lines required to connect the six points in Figure 4-4 so that the lines are drawn without lifting your pencil off the paper?

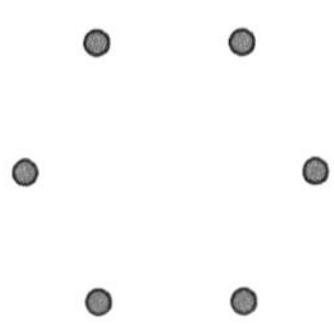

Figure 4-4

Without much thought, a typical response to this question is five lines. One could draw five lines in any one of the ways shown in Figure 4-5. But is this the least number of straight lines that can be used to connect these six points?

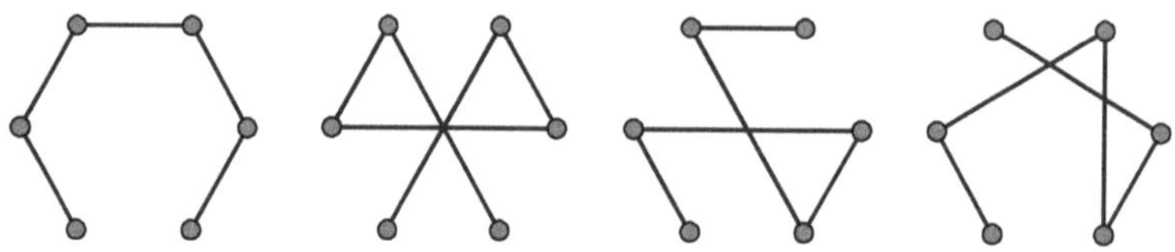

Figure 4-5

Solution: Logical thought can actually produce fewer lines to be used to connect the six points. The psychological mistake rests in the assumption that each line segment had to terminate at one of the points. As you can see in Figure 4-6, we were able to connect the six dots with four straight lines.

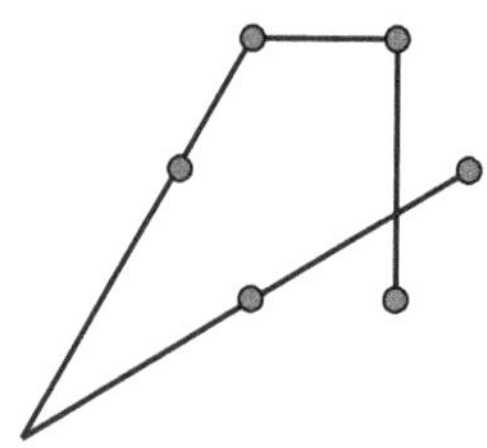

Figure 4-6

After abandoning this restriction of having line segments end at one of the given points, we can get an even better solution, namely, using three line segments, as we show in Figure 4-7.

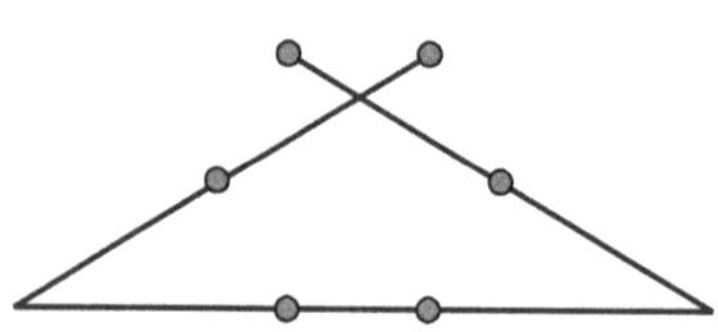

Figure 4-7

Our previous logical reasoning in this problem should be instructive for the next situation, where analogous logic should be used.

Problem 3: This time we are given nine dots, as shown in Figure 4-8, and asked to connect them with four straight lines without lifting your pencil off the paper until the points are joined.

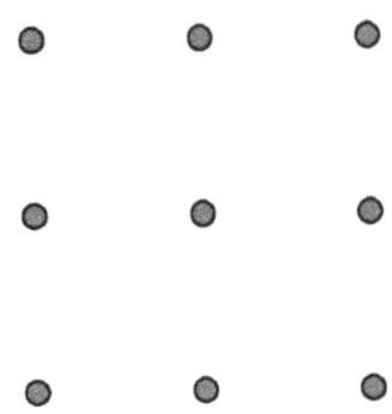

Figure 4-8

Solution: As we learned from the previous problem, a logical approach should not be limited to lines with endpoints at the given points. Our reasoning should enable you to arrive at solutions offered in Figure 4-9, where we need only four lines to connect all nine points.

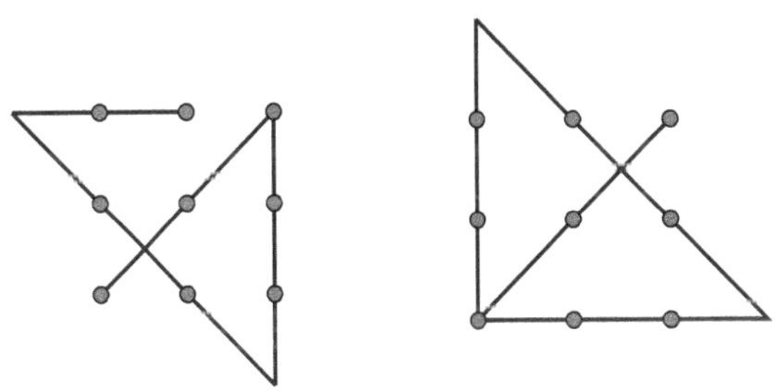

Figure 4-9

Problem 4: Now that logical reasoning has provided you with newfound skills, let's see if you can meet the following challenges. The first one is to connect the 12 dots shown in Figure 4-10 with as few as five straight lines, without lifting the pencil off the paper, and returning to the starting point.

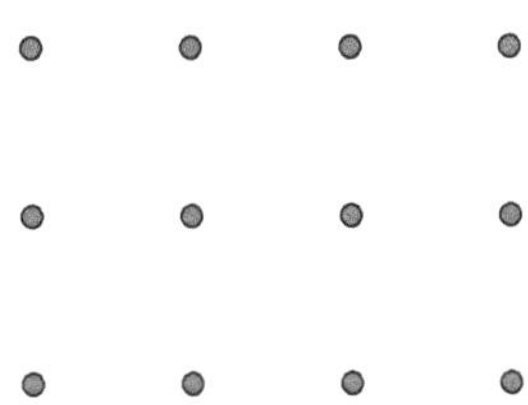

Figure 4-10

Solution: One solution is shown in Figure 4-11, but there may be others for the motivated reader to seek.

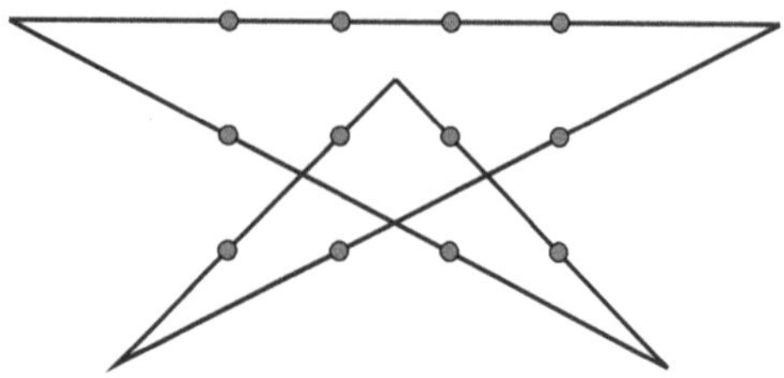

Figure 4-11

Problem 5: A further challenge is to connect the 25 dots, shown in Figure 4-12, using only eight straight lines and without lifting the pencil off the paper, but this time one does not have to begin and end at the same point.

Figure 4-12

Solution: Using nine lines would not be so difficult; however, with eight lines it becomes quite challenging. Nevertheless, we offer a solution shown in Figure 4-13.

These exercises, although somewhat challenging, provide reasonable training for logical thinking when one has to think beyond the traditional methods.

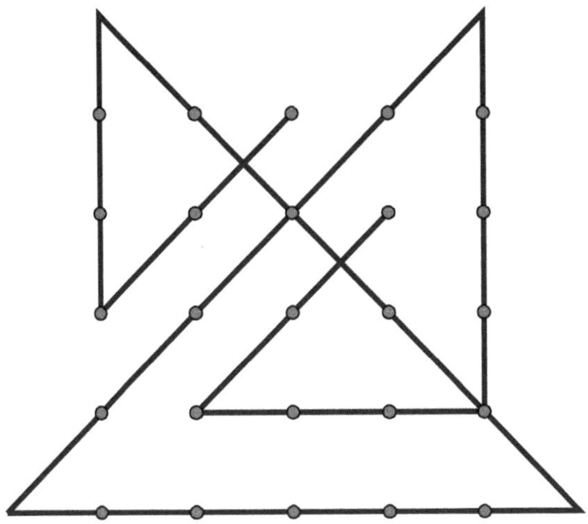

Figure 4-13

Problem 6: Lisa has an 11-liter can and a 5-liter can. How can she measure exactly 7 liters of water?

Solution: One might simply guess at the answer and keep pouring water back and forth between the two cans in an attempt to arrive at the correct answer, unsystematically guessing and testing. However, the problem can be solved in a more organized manner by working backwards. We need to end up with 7 liters in the 11-liter can, that is, removing 4 liters from the full 11-liter can. How can we measure 4 liters to remove from a full 11-liter can? (See Figure 4-14.)

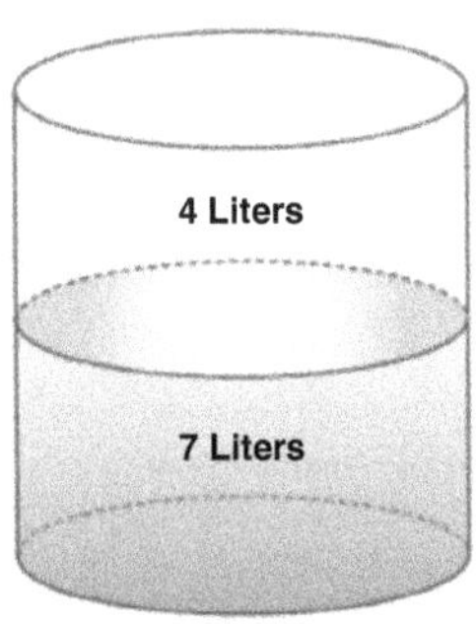

Figure 4-14

To remove 4 liters from the 11-liter can, we must leave 1 liter in the 5-liter can. Now, how can we obtain 1 liter in the 5-liter can? Fill the 11-liter can and pour from it twice into the 5-liter can, discarding the water. This leaves 1 liter in the 11-liter can. Pour the 1 liter into the 5-liter can. (See Figure 4-15.)

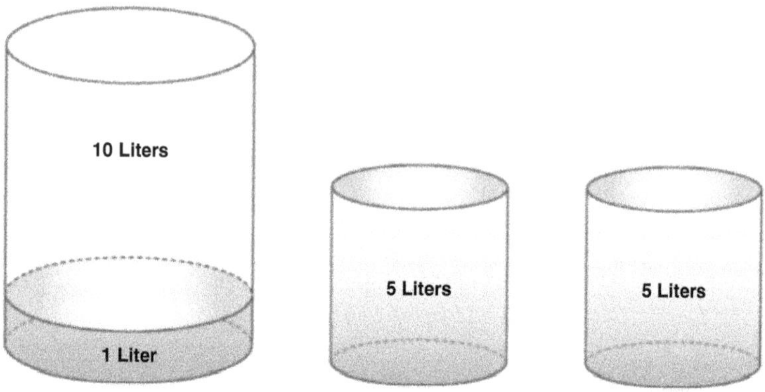

Figure 4-15

Now, fill the 11-liter can and pour off the 4 liters needed to fill the 5-liter can. This leaves the required 7 liters in the 11-liter can. (See Figure 4-16.)

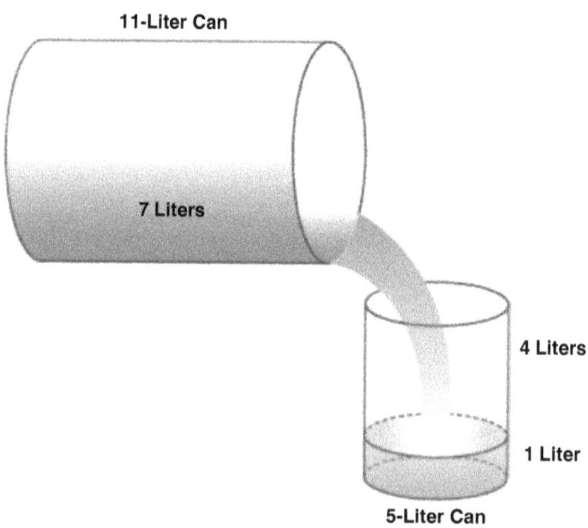

Figure 4-16

Note that problems of this sort do not always have a solution. That is, if you wish to construct additional problems of this sort, you must bear in mind that a solution only exists when the difference of multiples of the capacities of the two given cans can be made equal to the desired quantity. In this problem, 2(11) – 3(5) = 7.

Problem 7: Suppose you have nine coins, which all look the same and yet one coin is lighter than the other eight coins. Using a balance scale, how can you determine which is the lighter coin with only two weighings?

Solution: The challenge here is to come up with a way to do the weighings, while being limited to two tries. Begin by taking any three coins and weighing them against any other three coins. If the scale balances, then you will know that the lighter coin is among the remaining three coins that have not been weighed. Then take two of the three unweighed coins and place one on each end of the scale. If they balance, then the third coin—the one that has not yet been placed on the scale—is the lighter coin. If they don't balance, then the lighter coin will have revealed itself on the scale.

On the other hand, if the scale does not balance when the two groups of three coins are initially weighed, then the lighter group of three will contain the lighter coin. In that case, use the procedure above to determine the lighter coin amongst the three in the lighter group.

Problem 8: Bowling pins were originally set up by pin-boys and not by machine. Returning now to these old days, suppose the pin-boy set up the pins in the wrong direction. How can the direction of the pins shown in Figure 4-17 be reversed by moving only three pins?

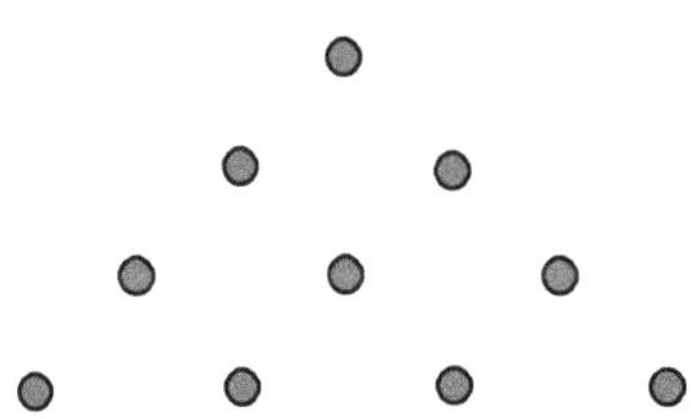

Figure 4-17

Solution: After multiple attempts to perform this task, the trickster might want to show how to move just three pins and have the direction of the triangle reversed, which can be seen in Figure 4-18.

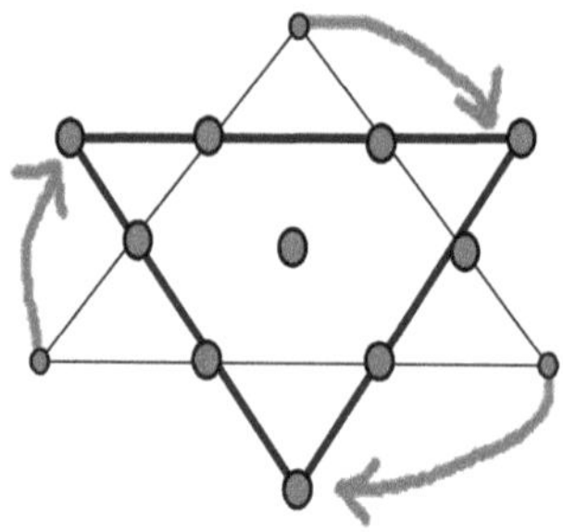

Figure 4-18

Problem 9: This challenge can be done by providing four chain pieces, each consisting of three links, as shown in Figure 4-19. One might also be able to solve it in a virtual sense. The challenge here is to show how these four chain pieces could be made into a circular chain by opening and closing, *at most*, three links.

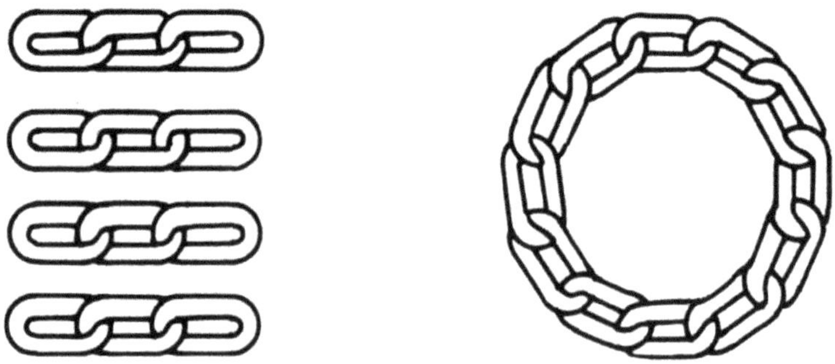

Figure 4-19

Solution: Typically, the first attempt at a solution involves opening the end link of one chain, joining it to the second chain to form a six-link chain, then opening and closing a link in the third chain and joining it to the six-link chain to form a nine-link chain. By opening and closing a link in the fourth chain and joining it to the nine-link chain, a 12-link chain is obtained, but it is *not* a circle. Thus, this typical attempt ends unsuccessfully. Most attempts usually involve other combinations of opening/closing one link of each of various chains to try to join them together to get the desired result, but this approach will not work.

Let's look at this from another point of view. Instead of continually trying to open and close *one* link of each chain piece, try opening *all* the links in one chain and using those links to connect the remaining three chain pieces together into the required circle chain. This method quickly gives us the sought-after solution.

Problem 10: Here is a challenge that will require a little thinking because of the wording involved. Suppose you are painting a square wall that measures 8ft × 8ft. When you are half done, you tell your friend that the remaining unpainted half of the wall is a square that measures from top-to-bottom 8 feet from the original top and 8 feet from side-to-side. What might the situation look like?

Solution: The challenge here lies in the orientation of the square, which most people, from a psychological point of view, do not expect at first glance. In Figure 4-20, we show the solution.

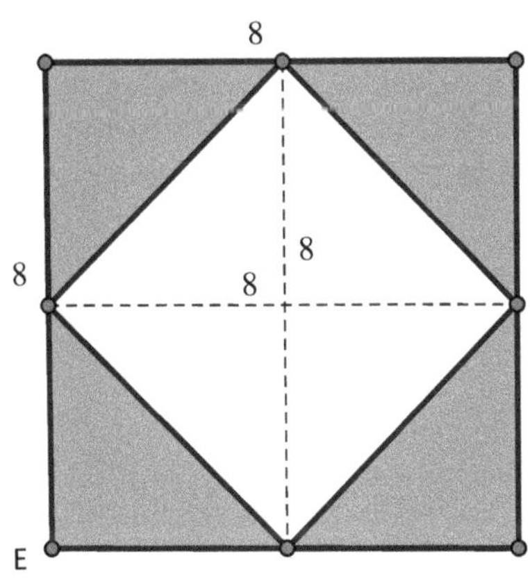

Figure 4-20

You will notice in Figure 4-20 that the distance from top-to-bottom and side-to-side is still 8 feet, which is what the task called for. It is probably quite surprising how this simple solution evades most people.

Problem 11: Here we have another challenge where an unsuspecting person will render the wrong answer and be surprised at the correct answer. A snail is at the bottom of a 10-foot well. During the day, it crawls up 3 feet. But at night it slides back 2 feet. So, in one day it accomplishes a distance of 1 foot in its quest to reach the top of the well. How many days will it take the snail get out of the well?

Solution: The typical response to this question is that the snail would exit after 10 days. However, at the end of the seventh day, the snail will have accomplished 7 feet of travel. Then on the morning of the eighth day the snail will crawl 3 feet and reach the top of the well, because the snail will not fall back that night after having reached the goal. In short, the snail will crawl out on the eighth day.

Problem 12: When a problem at first looks a bit daunting, and then a simple solution is presented, the general reaction is awe. We often wonder why we didn't think of that simple solution ourselves. Such problems that have a gee-whiz dramatic effect on us will likely help us with future (analogous) situations. Here is one such problem.

On a shelf in Danny's basement, there are three jars. One contains only nickels, one contains only dimes, and one contains a mixture of nickels and dimes. The three labels, "nickels," "dimes," and "mixed," fell off, and were all put back on the wrong jars (see Figure 4-21).

Figure 4-21

Without looking, Danny can select one coin from one of the mislabeled jars and then correctly label all three jars. From which jar should Danny select the coin?

Solution: One may reason that the "symmetry" of the problem dictates that whatever we can say about the jar mislabeled "nickels" could just as well have been said about the jar mislabeled "dimes." Thus, if Danny chooses a coin from either of these jars, the results would be the same.

You should, therefore, concentrate your considerations on what happens if he chooses from the jar mislabeled "mixed." Suppose Danny selects a nickel from the "mixed" jar. Since this jar is mislabeled, it cannot be the mixed jar and must be, in reality, the nickel jar. Since the jar marked dimes cannot really be dimes, it must be the "mixed" jar. This leaves the third jar to be the dimes jar. You are probably thinking how simple the problem is now that you have the solution. It does demonstrate a certain beauty of logical thinking.

Problem 13: A young fellow bought a football for $30. Needing money immediately, he was able to sell it for $40. Sometime later, eager to play football, he bought the ball back for $50. After he finished playing, he found someone who wanted to buy it for $60. Did this young fellow make a profit or have a loss in these transactions? And how much was this profit or loss?

Solution: This trick is confusing by its very nature and needs to be sorted out logically piece-by-piece. When this young fellow sold the football for the first time, he made a profit of $10. When he bought the football a second time, he made another profit of $10. Consequently, his profit for the entire set of transactions was $20. We should not be distracted from the difference in price that he bought the football on the two purchases, since that is relatively irrelevant.

Problem 14: We now embrace a visual surprise. How can four straight lines be drawn to partition the diagram consisting of three squares, shown in Figure 4-22, into four equal areas. Don't look to the solution to discover this trick until you have made a good effort to solve the problem. This will allow you to properly appreciate it.

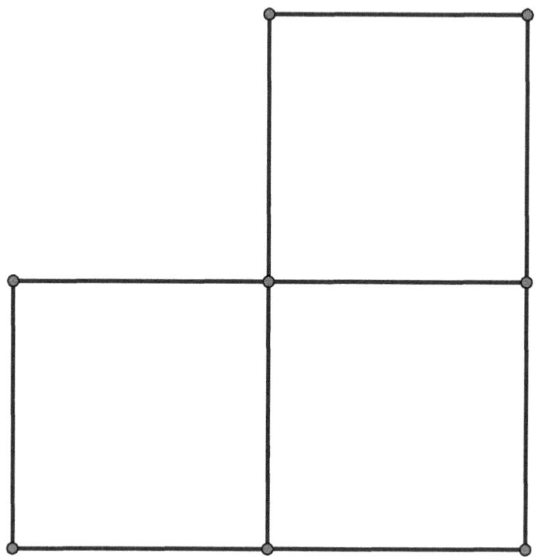

Figure 4-22

Solution: As we know, a picture is worth a thousand words; therefore, we shall go directly to the diagram shown in Figure 4-23. The shaded region is formed by selecting the midpoints of the various sides of the three original squares. Then draw four lines as shown in the Figure. Each of the three squares forming the shaded region is $\frac{1}{4}$ of the area of one of the three larger squares. Therefore, the remaining areas of each of the three original (larger) squares is $\frac{3}{4}$ of the original area of the square. Thus, we have four equal areas, each of which has an area $\frac{3}{4}$ of one of the original squares. The unusualness here lies in the unexpected construction.

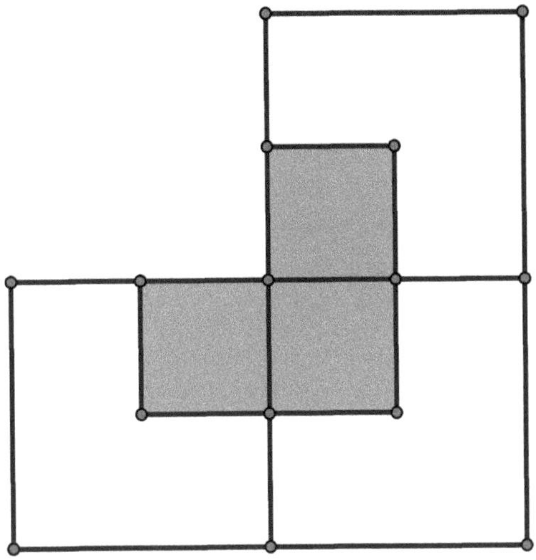

Figure 4-23

Problem 15: Two concentric circles are 10 units apart, as shown in Figure 4-24. What is the difference between the circumferences of the circles?

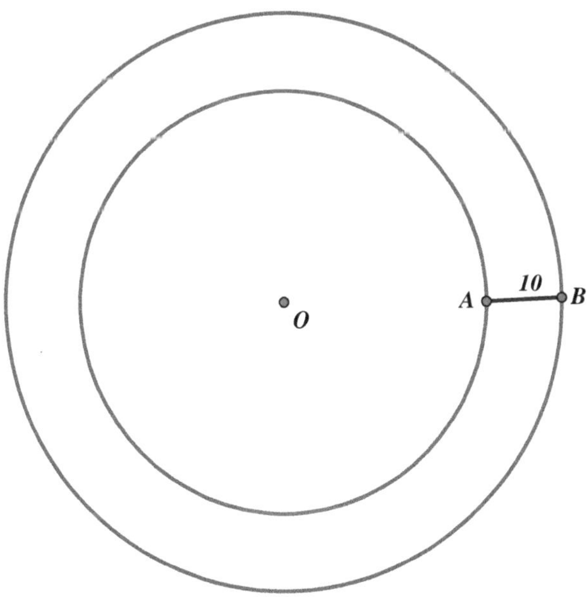

Figure 4-24

Solution: The traditional straightforward method for solving this problem is to find the length of the diameters of the two circles, calculate the circumference of each circle, and then find their difference. Because the lengths of the diameters are not given, the problem is a bit more complicated. Let d represent the diameter of the smaller circle. Then $d + 20$ is the diameter of the larger circle. The circumferences of the two circles will then be πd and $\pi(d + 20)$, respectively. Their difference is $\pi(d + 20) - \pi d = 20\pi$.

A more logical procedure is to consider an extreme case. Let the smaller of the two circles become smaller and smaller until it reaches an "extreme" and becomes a "point circle." In this case, it becomes the center of the larger circle. The distance between the two circles now becomes the radius of the larger circle. The difference between the lengths of the circumferences of the two circles is now merely the circumference of the larger circle, or 20π.

Although both procedures yield the same answer, notice how much more work is required for the traditional solution of actually taking the difference of the lengths of the circumferences of the two circles.

Problem 16: Imagine a metal hoop fitted tightly around the earth at the equator. Now cut the hoop and insert a piece exactly 10 feet long. When you hold the hoop in a position so that it is exactly concentric with the equator, how far above the surface of the earth will the hoop be?

Solution: One usually begins by drawing the situation as in Figure 4-25. In this figure, C is the original circumference of the hoop and r is the radius; C' is the circumference of the hoop with the 10 feet added, R is the new radius, and x is the distance (in feet) of the new hoop above the Earth's surface:

$$\text{(original circle)} \qquad C = 2\pi r$$

$$\text{(new circle)} \qquad C' = 2\pi R = C + 10$$

$$C + 10 = 2\pi(r + x)$$

$$C + 10 = 2\pi r + 2\pi x$$

$$10 = 2\pi x$$

$$\frac{5}{\pi} = x$$

$$1.6' = x.$$

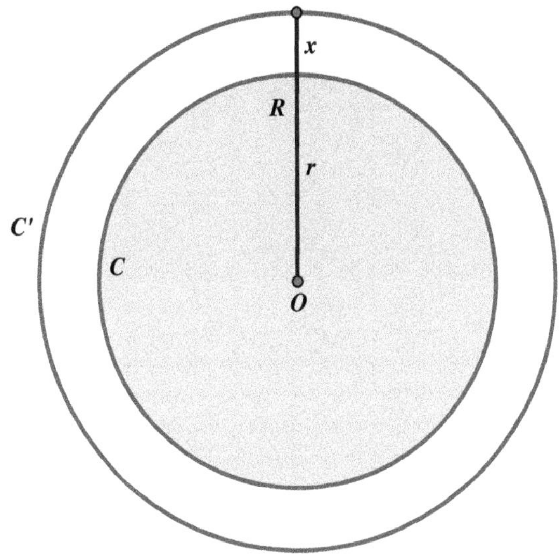

Figure 4-25

Using a bit of logic, we will solve this problem by considering the extreme case. Let the original circle shrink until r equals 0, as shown in Figure 4-26. Now, the original "hoop" has become point O, with $C = 0$.

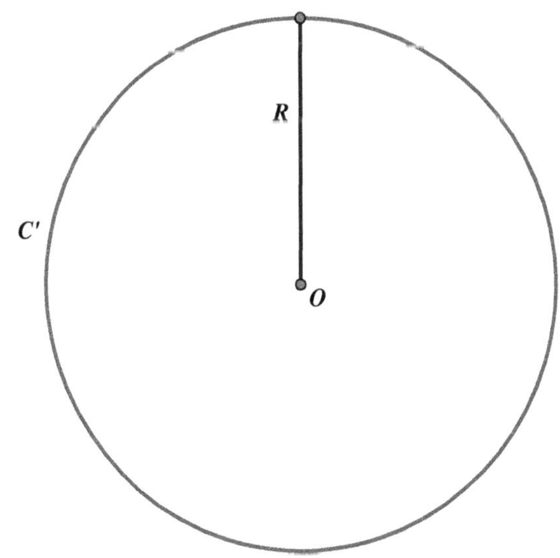

Figure 4-26

The distance we seek is the length of the radius of a circle with a circumference of 10 feet. Thus,

$$10 = 2\pi R$$
$$\frac{5}{\pi} = R$$
$$1.6 = R.$$

The distance above the surface is therefore 1.6 feet.

Problem 17: Here is another problem that lends itself to the use of the extreme case logic. In a drawer, there are 8 blue socks, 6 green socks, and 12 black socks. What is the smallest number of socks that must be taken from the drawer without looking at them to be certain of having 2 black socks?

Solution: This problem can be solved with extreme case logic, so let's set up the "worst-case scenario." This would have us pick all 8 blue socks and all 6 green socks before a single black sock is selected. The next two picks would have to be black socks. In this situation, it took 16 picks before we were *certain* of getting 2 black socks. Naturally, we might have achieved our goal of getting 2 black socks on our first two tries, but it was not assured, and it would have been highly unusual. Even if we picked 10 socks at random, we could not be certain that we had 2 black ones among them.

A simple extension of this problem is illuminating. Now that we know it would take 16 picks to be certain of getting 2 black socks, how many picks would be necessary to be certain of getting 4 black socks? If you know the answer immediately, you are a logical reasoning whiz! Yes, you only need two more picks, for a total of 18. You've already accounted for the worst case—picking all the socks of the other colors plus 2 black socks with your first 16 picks—so the next two picks can only provide you with 2 black socks. Using the logic of the extreme case can help define a given problem and then facilitate a solution.

Problem 18: In a drawer, there are 8 blue socks, 6 green socks, and 12 black socks. What is the smallest number of socks that must be taken from the drawer, without looking at the socks, to be certain of having 2 socks of the same color?

Solution: At first glance, this problem appears to be similar to problem 17, discussed above. However, there is a slight difference. In this case, we are looking for a matching pair of socks of *any* color. We now apply extreme-case logic, similar to that which we used previously. The worst-case scenario has us picking 1 blue sock, 1 green sock, and 1 black sock in our first three picks. Thus, the fourth sock must provide us with a matching pair, regardless of what color it is. The smallest number of socks to guarantee a matching pair is 4.

Problem 19: A rectangle has adjacent sides of lengths 10 units and 8 units, respectively. A second rectangle with adjacent sides of lengths 6 units and 4 units overlaps the first rectangle, as shown in Figure 4-27. What is the difference in area between the two nonoverlapping regions of the two rectangles?

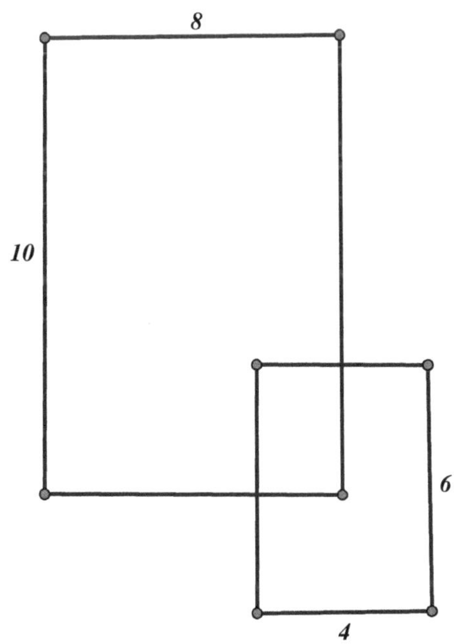

Figure 4-27

Solution: One may attempt to solve this problem by finding the area of each rectangle and then trying to somehow find the area of the common region. This would lead to frustration. Consider the logic of the extreme case. Whereas the area and placement of the region of overlap were not specified, we could assume an overlap of 0, as shown in Figure 4-28.

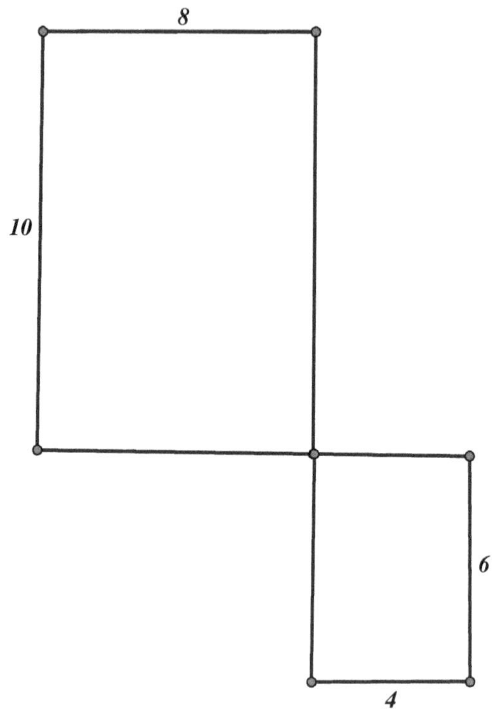

Figure 4-28

Now we see that the difference between the nonoverlapping areas is 80 square units – 24 square units = 56 square units. To reinforce our answer, let us examine two other cases. Consider rectangles that have a single unit of overlap, as shown in Figure 4-29.

The nonoverlapping difference is 79 square units – 23 square units = 56 square units. Let's consider another case, where the smaller rectangle is entirely enclosed within the larger, as shown in Figure 4-30.

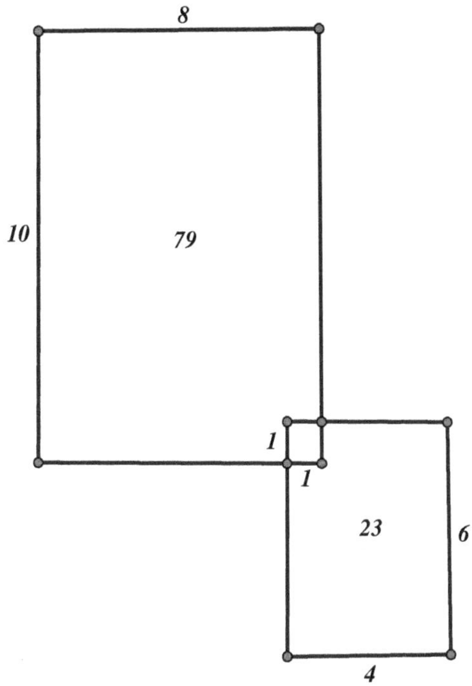

Figure 4-29

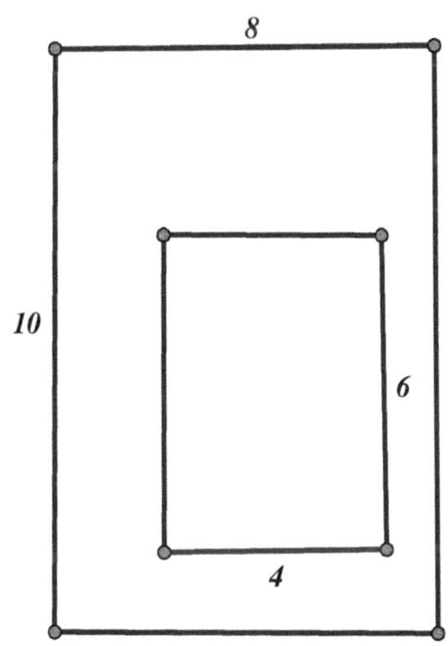

Figure 4-30

Again, the non-overlapping region is 56 square units. Once again, considering the extreme situations of a problem works as a logical approach. However, after observing the above extreme situations, one should notice that the difference of the areas has nothing to do with their placement. This is where logic overrides problem solving.

Problem 20: We offer here a challenge that clearly illustrates the value of considering extreme situations. There are two 1-gallon bottles, where one contains 1 quart of grape juice and the other contains 1 quart of apple juice. We take 1 tablespoon of grape juice and pour it into the apple juice. Then, we take 1 tablespoon of this new mixture (apple juice and grape juice) and pour it into the bottle of grape juice. The challenge here is to determine if there is more grape juice in the apple juice bottle or more apple juice in the grape juice bottle.

Solution: One can figure this out either through trial and error or by using logic. To consider the extreme case, let's take more than just a tablespoon. Instead, we remove an extremely large quantity: the entire 1 quart. That is, following the instructions given in the problem, we take the entire amount from one bottle (1 qt of grape juice) and pour it into the apple juice bottle. This mixture is now 50% apple juice and 50% grape juice. We then pour 1 quart of this mixture back into the grape juice bottle. The mixture is now the same in both bottles. Therefore, there is as much apple juice in the grape juice bottle as there is grape juice in the apple juice bottle!

We can also employ further logical thinking by considering the opposite extreme, where the spoon transporting the juice has a capacity of zero. In this case, the conclusion follows immediately: There is as much grape juice in the apple juice bottle as there is apple juice in the grape juice bottle, that is, zero.

Yet another solution to this problem uses alternative logical reasoning. With the first "transport" of juice there is only grape juice on the tablespoon. On the second "transport" of juice, the spoon carries a ratio of apple juice to grape juice proportional to the ratio in the apple juice bottle. To further firm up this logical reasoning, consider two containers, one with 12 white beans (called the white-bean container) and the other with 12 red beans (called the red-bean container). Suppose we move 6 white beans from the white-bean container to the red-bean container. This red-bean container will now have 12 red beans and 6 white beans, or 18 beans in total. From this red-bean container, we will remove 6 beans in the proportion of the contents (i.e., red:white = 2:1) back to the white-bean container. So, we will now move 4 red beans and 2 white beans, bringing the contents to the white-bean container to 8 white beans (6 + 2) and 4 red beans, while the red-bean container now remains with 4 white beans and 8 red beans. In other words, there are as many red beans (4) in the white-bean container as there are white beans (4) in the red-bean container.

Problem 21: A single elimination basketball tournament (one loss and the team is eliminated) has 25 teams competing. How many games must be played until there is a single tournament champion?

Solution: Typically, one might simulate the tournament, which does not employ logical thinking. This procedure would begin by taking, for example, two groups of 12 teams playing each other during the first round, with one team drawing a bye (**12 games played**). After the first round, 12 teams will have been eliminated, leaving 12 teams and the one team that drew a bye remaining in the tournament. In the next round, of these 13 teams, 6 teams will play another 6 teams, leaving 6 winners and the one team that drew a bye (**6 games played**). In round three, of the 7 teams remaining, 3 will play another 3, leaving 3 winners, plus the team that drew a bye (**3 games played**). In round four, the 4 remaining teams will play each other (**2 games played**), leaving 2 remaining teams that will play each other for the championship (**1 game played**). They then count the number of games played to get a champion team, 12 + 6 + 3 + 2 + 1 = 24. This is a perfectly legitimate problem-solving technique that does not use logic, but it's clearly not the most elegant solution.

Let us now consider a simple logical solution, one that most people do not realize on their first attempt, as its focus only is on the losers and not on the winners. A logical question to ask is, "How many losers must there be in the tournament with 25 teams in order for there to be one winner?" The answer is simple: 24 losers. Then how many games must be played to get 24 losers? Naturally, 24 games. So there you have the answer, very simply done with a little logical reasoning.

There is an interesting alternative to the above solution that can be put as follows: suppose that of the 25 teams, there is clearly one that is vastly superior to all the others. We could have each of the remaining 24 teams play this superior team and, of course, lose each game. Here again, you can see that only 24 games are required to get the champion, in this case, the predetermined superior team.

Problem 22: At 5 o'clock, a clock strikes 5 chimes in 5 seconds. How long will it take the same clock at the same rate to strike 10 chimes at 10 o'clock? (Assume that the chime itself takes no time.) Hint: the answer is *not* 10 seconds.

Solution: It would be logically efficient to draw the situation to see exactly what is taking place, as shown in Figure 4-31, where each circle represents a chime. Here, the total time is 5 seconds, where the chimes themselves require no time, and there are four intervals between chimes. Therefore, each interval must take $\frac{5}{4}$ seconds. At 10 o'clock, there will be *nine* intervals between the chimes, where each interval takes $\frac{5}{4}$ seconds. Therefore, the clock striking at 10:00 will take $9\times\frac{5}{4}=11\frac{1}{4}$ seconds.

Figure 4-31

Problem 23: Using two egg timers, one of which runs for exactly 7 minutes and the other for exactly 11 minutes, show how to cook an egg for exactly 15 minutes.

Solution: To solve the problem, we will use a logical strategy of working backwards. Consider the end result, 15 minutes. This can be obtained from consecutive timings of 11 and 4 minutes. We have an 11-minute timer, but the 4 minutes presents a possible difficulty. How can we measure exactly 4 minutes? Since $11 - 7 = 4$, we begin the timing by starting both timers simultaneously. When the 7-minute timer rings, begin cooking the egg. This leaves exactly 4 minutes on the 11-minute timer. After the 4 minutes have elapsed, restart the 11-minute timer. When it finally rings, the egg will have cooked for exactly 15 minutes. Logical thinking solves the problem.

Problem 24: In Figure 4-32, unit square *ABCD* has 2 quarter circles, *AFCB* and *AECD*, that have *B* and *D* as their respective centers in the square's sides as their radii. The challenge is to find the area of the "football" shaped region *AECF*.

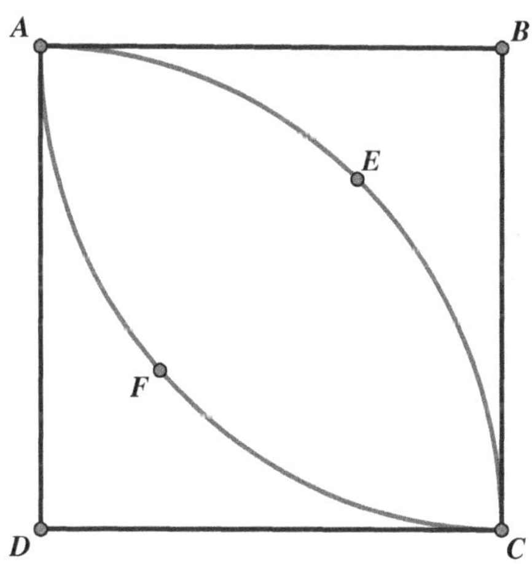

Figure 4-32

Solution: To find the area of unusual-shaped figures, one typically divides the figure into portions whose areas can be found, and then add or subtract as needed. Thus, most would begin by drawing the diagonal *AC*. While this can lead to a satisfactory solution, it is not the most logical approach. Stepping back from the traditional approach and using logic has us partitioning the unit square *ABCD* into three regions, as shown in Figure 4-33: region 1, bounded by *ADCF*; the football-shaped region 2, bounded by *AECF*; and region 3, bounded by *ABCE*. We can add the areas of the two sectors *AECD* (regions 1 and 2) and *AFCB* (regions 2 and 3). As they are each a quarter of the area of the unit circle, the result is one-half the area of the circle, or $\frac{\pi}{2}$ = (region 1 + region 2) + (region 2 + region 3). Subtracting the area of square *ABCD* (which consists of regions 1, 2, and 3) leaves the area of region 2, which was used twice when adding *AECD* and *AFCB* and only once when subtracting *ABCD*:

$$2\left(\frac{1}{4}\pi r^2\right)-r^2=\frac{\pi r^2}{2}-r^2.$$

We can let $r=1$, then we get $\dfrac{\pi}{2}-1$.

Thus, the area of *AECF* is $\frac{\pi}{2}-1$.

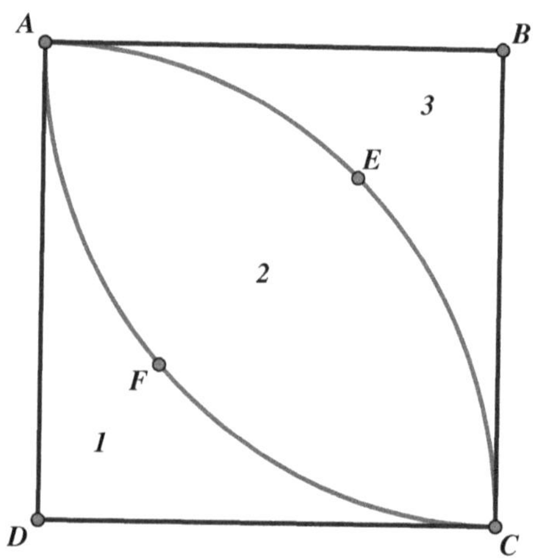

Figure 4-33

Problem 25: While the famous mathematician Carl Friedrich Gauss was in elementary school, he exhibited a wonderful logic that surprised even his teacher. The teacher wanted to occupy the class so that he could do some clerical chores at his desk. He asked the class to add the numbers from 1 to 100 in half an hour.

Solution: As the class began to get the sum of $1 + 2 + 3 + 4 + 5 + \cdots + 97 + 98 + 99 + 100$, Carl raised his hand with the result in less than a minute. Of course, the teacher ignored him. But at the end of half-hour, Carl was probably the only child to get the right answer. What he did was think logically. He noticed that instead of adding the numbers in consecutive order, as the rest of the class did, he found the sums as: $1 + 100 = 101$, $2 + 99 = 101$, $3 + 98 = 101$, etc. He concluded that he only had to multiply $50 \times 101 = 5{,}050$.

Problem 26: The divisors of 360 add up to 1170. What is the sum of the reciprocals of the divisors of 360?

Solution: This problem lends itself to a very clever logical technique. First, let's consider how one would traditionally approach the problem. The most obvious solution would be to find all the divisors of 360, take their reciprocals, and then add. The divisors of 360 are 1, 2, 3, 4, 5, 6, 8, 9, ... ,120, 180, 360. Their reciprocals are $\frac{1}{1}, \frac{1}{2}, \frac{1}{3}, \frac{1}{4}, \frac{1}{5}, \frac{1}{6}, \frac{1}{8}, \frac{1}{9}, \ldots, \frac{1}{120}, \frac{1}{180}, \frac{1}{360}$. Then we convert all the fractions to their equivalents with the common denominator of 360, add them up, and, barring any computational errors, we get the correct answer.

A far more logical approach would be to see if there is a pattern between the sum of the divisors and the sum of their reciprocals. Let's examine a simpler analogous problem using the number 12 instead of the number 360 to see if there is a pattern that could be helpful. The divisors of 12 are 1, 2, 3, 4, 6, and 12. Their sum is $1 + 2 + 3 + 4 + 6 + 12 = 28$. Now let's find the sum of the reciprocals of these factors:

$$\frac{1}{1} + \frac{1}{2} + \frac{1}{3} + \frac{1}{4} + \frac{1}{6} + \frac{1}{12} = \frac{28}{12}.$$

Aha! The numerator of the fraction is the sum of the divisors, while the denominator is the number we are working with. Now we can solve our original problem. The sum of the factors of 360 is 1170. Thus, the sum of the reciprocals of the factors must be $\frac{1170}{360}$.

This logical approach can be quite valuable for analogous problems.

Problem 27: Find all pairs of prime numbers whose sum equals 999.

Solution: One might begin to solve this problem by taking a list of prime numbers and trying various pairs to see if they obtain a sum of 999. This is obviously very tedious and time consuming, and one would never be quite certain that they had considered all possible prime number pairs.

Let's use our logical reasoning to solve this problem. In order to obtain an odd sum from two numbers (prime or otherwise), exactly one of the numbers must be even. Since there is only one even prime, namely, 2, there can be only one pair of primes whose sum is 999, and that pair is 2 and 997.

Problem 28: A student sold 51 theater tickets in exactly three days. Each day she sold 2 more theater tickets than she sold on the previous day. How many theater tickets did she sell on each day?

Solution:

Most people might approach the problem from an algebraic point of view:

x = the number of theater tickets sold on the first day
$x + 2$ = the number of theater tickets sold on the second day
$x + 4$ = the number of theater tickets sold on the third day.

Therefore, the sum of tickets sold can be represented as follows:

$$\begin{aligned} x+(x+2)+(x+4)&=51\\ 3x+6&=51\\ 3x&=45\\ x&=15. \end{aligned}$$

She sold 15 theater tickets the first day, 17 theater tickets the second day, and 19 theater tickets the third day.

Now, let's look at this problem from the point of view of logical reasoning. Since she sold 51 theater tickets in three days, she sold an average of 17 theater tickets per day. Since the difference between the number sold on each day is a constant, the 17 represents the number sold on the "middle" day. Thus, on the day previous, she sold 17 – 2 or 15 theater tickets, and on the following day she sold 17 + 2, or 19 theater tickets.

Problem 29: Geometry problems can also be solved by using simple logic. Consider circle *O*, where arc *DE* is 90°, as shown in Figure 4-34. The base *BC* of isosceles triangle *ABC* ($AB = AC$) contains the center of the circle *O*. What is the relationship between the area of triangle *ABC* and the area of triangle *DOE*?

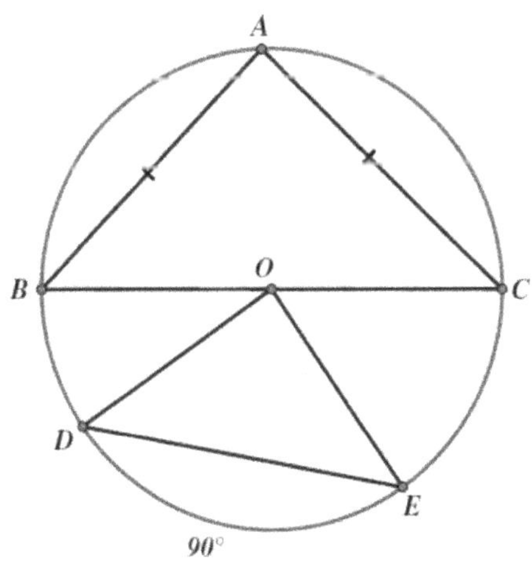

Figure 4-34

Solution: There are many ways to solve this problem. However, simple logic would have us rotate triangle *DOE* so that points *B* and *D* coincide. We show this in Figure 4-35. It should then be obvious that triangle *DOE* is half the size of triangle *ABC*. Simple logic has saved us from calculating the areas of these geometric configurations.

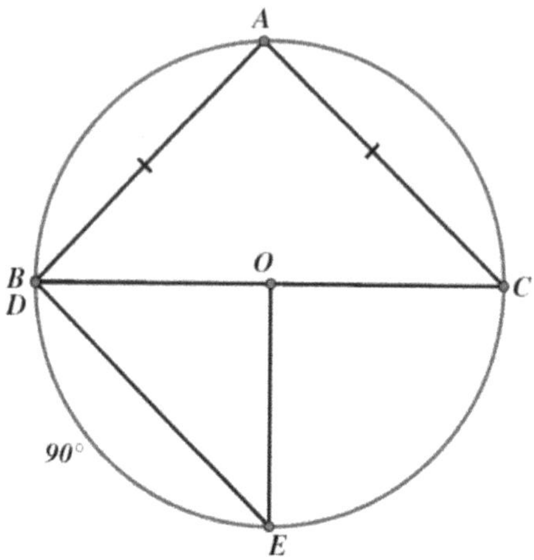

Figure 4-35

Problem 30: Two men are sorting out a financial question. Max owes Jack 4 dollars and Jack owes Max 5 dollars. Max gives Jack 6 dollars. What needs to be done to cancel the debt?

Solution: We embark on simple logical reasoning. When Max gives Jack 6 dollars, then Jack owes Max 6 dollars more than before, that is, he owes him 11 dollars. But since Max owes Jack 4 dollars, Jack then only owes Max 7 dollars. To cancel the debt, Jack must give Max 7 dollars.

Problem 31: This problem usually fools most people, unless logical reason is used. If 5 men can plant 5 trees in 5 days, how many men will it take the to plant 50 trees in 50 days?

Solution: This problem is deceptive. One assumes that the answer must be consistent with the original information, and so one gives the answer of 50 men. Now let's examine this logically. Since we know that 5 men can plant 5 trees in 5 days, then we can assume that 1 man can plant 1 tree in 5 days. If we multiply that by 10, we then have 1 man planting 10 trees in 50 days. When we multiply that by 5, we find that 5 men can plant 50 trees in 50 days, which answers our original question.

Problem 32: **S**uppose your friend drives from home to work at 30 mph due to heavy traffic. He then drives the same route back from work to home at 60 mph. What was his average speed for the entire round-trip? Most people would respond by taking the average of 30 and 60, as

$$\frac{30+60}{2}=45,$$

but this is incorrect. This is where logical reasoning plays a key role, which you will see with the solution.

Solution: First, the reason for 45 being a wrong answer is that the driver spent *twice as much* time driving at 30 mph as he did driving at 60 mph. Therefore, the two rates of speed cannot be given the same weight. The correct response would be to give the appropriate weights based on time spent, so that the correct average speed would be

$$\frac{30+30+60}{3}=40\text{mph}.$$

This is far more logical.

For those not convinced by this argument, try something a bit closer to "home." Say a student scored 100% on nine of ten tests in a semester, and on one test scored only 50%. What grade does she deserve? Would it be fair to assume that this student's performance for the term should be 75%, that is,

$$\frac{100+50}{2}=75?$$

This outrageous suggestion dissuades us from weighing the two scores in consideration equally. The 100% score was achieved nine times, and the 50% score was achieved only once. Therefore, scores of

100% must be weighed appropriately. A proper calculation of the student's average would to be

$$\frac{9(100)+50}{10}=95.$$

This clearly appears more just!

An astute reader may now ask "what happens if the rates to be averaged are not multiples of one another?" For the speed problem above, one could find the time "going" and the time "returning" to get the total time, and then with the total distance calculate the total rate, which is, in fact, the average rate.

A more efficient way to solve this problem enables us to employ uncommon concepts in mathematics. We are going to introduce a concept called the *Harmonic Mean,* which is the mean of a harmonic sequence, that is, a sequence of numbers whose reciprocals form an arithmetic sequence, one with common differences. The name "harmonic" may come from the fact that one such harmonic sequence is

$$\frac{1}{2},\frac{1}{3},\frac{1}{4},\frac{1}{5},\frac{1}{6},\frac{1}{7},\frac{1}{8},$$

and if one takes guitar strings of these relative lengths and strums them together, a harmonious sound results.

This frequently misunderstood mean (or average) usually causes confusion, but once we understand that we are finding the average of rates (i.e. the harmonic mean), then we have a useful formula for calculating the harmonic mean for rates over the same base. In the above situation, the rates were for the same distance (round trip legs).

The harmonic mean for two rates a and b is

$$\frac{2ab}{a+b}.$$

For three rates, a, b, and c, the harmonic mean is

$$\frac{3abc}{ab+bc+ac}.$$

You can see the pattern evolving, so that for four rates, the harmonic mean is

$$\frac{4abcd}{abc+abd+acd+bcd}.$$

Now applying this formula to the above speed problem gives us

$$\frac{2\times30\times60}{30+60}=\frac{3600}{90}=40.$$

This topic is not only useful, but also serves to sensitize one to the notion of weighted averages, a very important concept to remember.

Problem 33: We are given a chessboard and 32 dominos, each of which is the exact size of two of the squares on the chessboard. Can you show how 31 of these dominos can cover the chessboard when a pair of opposite squares have been removed, as shown in Figure 4-36?

Figure 4-36

Solution: As soon as the question is posed, most people try covering the board with various arrangements of dominos. This may be done with actual tiles, or by drawing the board on grid paper and then shading adjacent squares two at a time. Before long, frustration begins to set in, since this approach cannot be successful. Here, the solution lies back in the original question. A careful reading of the question reveals that it does not say to do this tile covering; it asks *if it can* be done. Yet, because of the way we have been trained to respond to problems, the question is often misread and interpreted as "do it."

A bit of logical insight helps. Ask the question: "When a domino tile is placed on the chessboard, what kind of squares are covered?" A black square and a white square must be covered by each domino placed on the chessboard. Is there an equal number of black and white squares on the truncated chessboard? No! There are two fewer black squares than white squares. Therefore, it is impossible to cover the truncated chessboard with the 31 domino tiles, since there must be an equal number of black and white squares. Thus, the question has then been answered.

Chapter 5

Interesting Aspects of Mathematics

A History of the Fibonacci Numbers

In the previous chapters we exhibited aspects of mathematics that are easy to understand and appreciate. They give the reader the opportunity to appreciate mathematics from a non-scholastic standpoint. In that spirit, this chapter will present mathematics through a cultural lens. We shall begin with an introduction to the most famous numbers in the realm of mathematics. These numbers are involved in just about every aspect of our experiences: arithmetic, algebra, geometry, architecture, biology, botany, etc. We begin with the history of the *Fibonacci numbers*.

With the dawn of the 13th century, Europe began to wake from the long sleep of the Middle Ages and perceive faint glimmers of the coming Renaissance. The mists rose slowly as the forces of change impelled scholars, crusaders, artists, and merchants to take their tentative steps into the future. Nowhere were these stirrings more evident than in the great mercantile cities of Italy. By the end of the 13th century, Marco Polo had journeyed the Silk Road to reach China, Giotto di Bondone had altered the course of European painting and freed it from Byzantine conventions, and the mathematician Leonardo Pisano, best known as Fibonacci, had changed forever Western methods of calculation, which facilitated trade and the exchange of currency (Figure 5.1).

Figure 5-1 Leonardo of Pisa (Fibonacci)

Leonardo Pisano, or Leonardo of Pisa, is known today as Fibonacci,[1] a name derived from either the Latin *filius Bonacci*, meaning "a son of Bonacci," or from the Latin *de filiis Bonacci*, meaning "family of Bonacci." He was born to Guglielmo Bonacci and his wife around 1170 in the port city of Pisa, Italy. In 1192, Guglielmo Bonacci became a public clerk for the Republic of Pisa, and was stationed in the customs house in the Pisan colony of Bugia (later Bougie, and today Bejaia, Algeria) on the Barbary Coast of Africa. Shortly after his arrival he sent for his son to join him so that Leonardo could learn the skill of calculating and become a merchant. This calculating ability was a significant asset, since each city, republic, and kingdom had its own currency, and traders had to calculate monies to trade. This meant determining exchange rates on a daily basis. It was in Bugia that Fibonacci first became acquainted with the "nine Indian figures," as he called the Hindu numerals, and "the sign 0 which the Arabs call zephyr," which were completely unknown in Europe. Fibonacci declared his fascination with these numerals, and the methods of calculation they enabled, in the only source we have about his life story: the prologue to his book *Liber Abaci*. During his time away from Pisa, he received instruction from a Muslim teacher who introduced him to a book on algebra entitled *Hisâb al-Jabr w'al-Muqabâlah*[2] by the

[1] It is unclear who first used the name Fibonacci; however, it seems to have been either Giovanni Gabriello Grimaldi (1757–1837) around 1790, or Pietro Cossali (1748–1815).

[2] The word "algebra" comes from the title of this book.

Persian mathematicia al-Khowarizmi (ca. 780 – ca. 850), which also influenced him. When he returned to Pisa around the turn of the century, Fibonacci began to write about calculation methods using the Indian numerals for commercial applications in his *Liber Abaci*, which means "book on calculation."[3] The book consists primarily of algebraic problems and "real world" problems that require more abstract mathematics. Fibonacci wanted to introduce these newfound techniques to his countrymen.

Before the printing press, all books had to be written by the hand of scribes, and if a copy was to be made that, too, had to be handwritten. Therefore, we are fortunate to still have copies of *Liber Abaci*, which first appeared in 1202 and was later revised in 1228.[4] In 1225, he wrote the books *Flos* ("on flowers") and *Liber Quadratorum* ("book of squares"). The latter distinguished Fibonacci as a talented mathematician, ranking very high amongst number theorists. As a strong patron of science and the arts, Emperor Frederick II (1194–1250) became aware of Fibonacci's work through the scholars at his court. Johannes of Palermo, another member of Frederick II's court, presented a number of problems as challenges to the great mathematician Fibonacci. Fibonacci solved some of these problems and provided their solutions in *Flos*, which he sent to Frederick II.

One of these problems, taken from Omar Khayyam's book on algebra, was to solve the equation $x^3 + 2x^2 + 10x = 20$. Fibonacci knew that this was not solvable with the numerical system then in place, that is, Roman numerals. He pointed out that the answer was not an integer, nor a fraction, nor the square root of a fraction. Without any explanation, Fibonacci provided an approximate solution in the form of a sexagesimal (that is, a number in base 60) number as 1.22.7.42.33.4.40, which equals

$$1+\frac{22}{60}+\frac{7}{60^2}+\frac{42}{60^3}+\frac{33}{60^4}+\frac{4}{60^5}+\frac{40}{60^6}.$$

However, with today's computer algebra system we can get the proper solution, which is by no means trivial. It is

---

[3] Despite the false cognate, it is not about the abacus.

[4] Several copies of the second edition from 1228 still exist. *Liber Abaci* was first printed in modern times by Baldassarre Boncompagni in 1857. The first English edition was produced by Laurence E. Sigler (New York: Springer-Verlag, 2002), based upon Boncompagni's 1857 printing.

$$x = -\sqrt[3]{\frac{2\sqrt{3930}}{9} - \frac{352}{27}} + \sqrt[3]{\frac{2\sqrt{3930}}{9} - \frac{352}{27}} - \frac{2}{3} \approx 1.3688081075.$$

We can explore another of these challenging problems here, since it only requires some elementary algebra. However, although these methods may seem elementary to us, they were hardly known at the time of Fibonacci, and so this was considered a real challenge. The problem was to find the square number that remains a square number when increased or decreased by 5.

Fibonacci found the number $\frac{41}{12}$ as his solution to the problem. To check this, we must add and subtract 5 and see if the result is still a perfect square.

$$\left(\frac{41}{12}\right)^2 + 5 = \frac{1681}{144} + \frac{720}{144} = \frac{2401}{144} = \left(\frac{49}{12}\right)^2$$
$$\left(\frac{41}{12}\right)^2 - 5 = \frac{1681}{144} - \frac{720}{144} = \frac{961}{144} = \left(\frac{31}{12}\right)^2.$$

We have thus shown that $\frac{41}{12}$ meets the criteria set out in the problem.

The last mention of Fibonacci was in 1240, when the Republic of Pisa honored him with a lifetime salary for his services to the people, whom he advised on matters of accounting, often pro bono.

The Book, Liber Abaci

Although Fibonacci wrote several books, the one we will focus on is *Liber abaci*. This extensive volume is full of very interesting problems. Based on the arithmetic and algebra that Fibonacci had accumulated during his travels, it was widely copied and imitated. Most importantly, as noted, *Liber Abaci* introduced into Europe Arabic numerals and the Hindu-Arabic place-valued decimal system. The book was widely used for the better part of the next two centuries—a best seller!

Fibonacci begins *Liber Abaci* with the following:

"The nine Indian[5] figures are: 9 8 7 6 5 4 3 2 1.[6]

---

[5] Fibonacci used the term "Indian figures" for the Hindu numerals.

[6] It is assumed that Fibonacci wrote the numerals in order from right to left, since he took them from the Arabs who write in this direction.

> With these nine figures, and with the sign 0, which the Arabs call zephyr, any number whatsoever is written, as demonstrated below. A number is a sum of units, and through the addition of them the number increases by steps without end. First, one composes those numbers, which are from one to ten. Second, from the tens are made those numbers, which are from ten up to one hundred. Third, from the hundreds are made those numbers, which are from one hundred up to one thousand. ... and thus, by an unending sequence of steps, any number whatsoever is constructed by joining the preceding numbers. The first place in the writing of the numbers is at the right. The second follows the first to the left."

Despite their relative facility, these numerals were not widely accepted by merchants, who were suspicious of those who knew how to use them for fear of being cheated. It took over 50 years before these numerals began to be used in Europe.

Interestingly, *Liber Abaci* also contains simultaneous linear equations. Many of the problems that Fibonacci considers, however, were similar to those appearing in Arab sources. This does not detract from the value of the book, since it is the collection of solutions to these problems that makes the major contribution to the development of mathematics. As a matter of fact, a number of mathematical terms—common in use today—were first introduced in *Liber Abaci*. Fibonacci referred to *factus ex multiplicatione*,[7] and from this we speak of the "factors of a number" or the "factors of a multiplication." Another example of words Fibonacci introduced into our mathematical vocabulary are the words "numerator" and "denominator."

The second section of *Liber Abaci* includes a large collection of problems intended for merchants. They relate to the price of goods, how to convert between the various currencies in use in Mediterranean countries, how to calculate profit on transactions, and some problems that may have originated in China. To circumvent the church's ban on charging interest on loans, Fibonacci devised a way to hide the interest in a higher initial sum than the actual loan and based his calculations on compound interest.

The book *Liber Abaci* is of interest to us not only because it was the first publication in Western culture to use the Hindu numerals

---

[7] Smith, David Eugene. *History of Mathematics*, vol. 2, (New York: Dover, 1958), p. 105.

instead of the clumsy Roman numerals, or because Fibonacci was the first to use a horizontal fraction bar, but because it casually includes a recreational mathematics problem in Chapter 12 that has made Fibonacci famous for posterity. This is the problem dealing with the regeneration of rabbits.

The Rabbit Problem

Figure 5-2 shows how the problem was stated (with the marginal note included):

Beginning	1
First	2
Second	3
Third	5
Fourth	8
Fifth	13
Sixth	21
Seventh	34
Eighth	55
Ninth	89
Tenth	144
Eleventh	233
Twelfth	377

"A certain man had one pair of rabbits together in a certain enclosed place, and one wishes to know how many are created from the pair in one year when it is the nature of them in a single month to bear another pair, and in the second month those born to bear also. Because the above written pair in the first month bore, you will double it; there will be two pairs in one month. One of these, namely the first, bears in the second month, and thus there are in the second month 3 pairs; of these in one month two are pregnant and in the third month 2 pairs of rabbits are born and thus there are 5 pairs in the month; in this month 3 pairs are pregnant and in the fourth month there are 8 pairs, of which 5 pairs bear another 5 pairs; these are added to the 8 pairs making 13 pairs in the fifth month; these 5 pairs that are born in this month do not mate in this month, but another 8 pairs are pregnant, and thus there are in the sixth month 21 pairs; to these are added the 13 pairs that are born in the seventh month; there will be 34 pairs in this month; to this are added the 21 pairs that are born in the eighth month; there will be 55 pairs in this month; to these are added the 34 pairs that are born in the ninth month; there will be 89 pairs in this month; to these are added again the 55 pairs that are both in the tenth month; there will be 144 pairs in this month; to these are added again the 89 pairs that are born in the eleventh month; there will be 233 pairs in this month. To these are still added the 144 pairs that are born in the last month; there will be 377 pairs and this many pairs are produced from the above-written pair in the mentioned place at the end of one year.

You can indeed see in the margin how we operated, namely that we added the first number to the second, namely the 1 to the 2, and the second to the third and the third to the fourth and the fourth to the fifth, and thus one after another until we added the tenth to the eleventh, namely the 144 to the 233, and we had the above-written sum of rabbits, namely 377 and thus you can in order find it for an unending number of months."[8]

Figure 5-2

[8] Translation from L. Sigler (2002).

To see how this problem would look on a monthly basis, consider the chart in Figure 5-3, where we assume that a pair of baby rabbits (B) mature in one month to become offspring-producing adults (A).

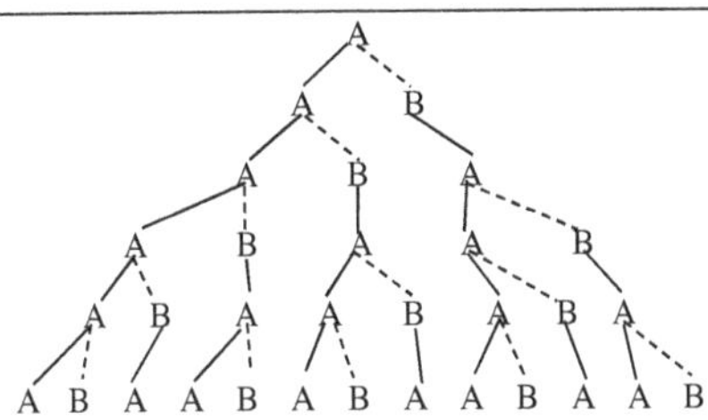

Month	Pairs	No. of Pairs of Adults (A)	No. of Pairs of Babies (B)	Total Pairs
Jan. 1		1	0	1
Feb. 1		1	1	2
Mar. 1		2	1	3
Apr. 1		3	2	5
May 1		5	3	8
June 1		8	5	13
July 1		13	8	21
Aug. 1		21	13	34
Sept. 1		34	21	55
Oct. 1		55	34	89
Nov. 1		89	55	144
Dec. 1		144	89	233
Jan. 1		233	144	377

Figure 5-3

This problem generated the sequence of numbers

$$1, 1, 2, 3, 5, 8, 13, 21, 34, 55, 89, 144, 233, 377, \ldots,$$

which are known today as the *Fibonacci numbers*. At first glance, there is nothing spectacular about these numbers beyond the relationship that would allow us to generate additional numbers of the sequence quite easily. We notice that every number in the sequence (after the first two) is the sum of the two preceding numbers. The Fibonacci sequence can be written in such a way that its recursive definition becomes clear:

$$\begin{array}{l}
1\\
1\\
1+1=2\\
1+2=3\\
2+3=5\\
3+5=8\\
5+8=13\\
8+13=21\\
13+21=34\\
21+34=55\\
34+55=89\\
55+89=144\\
89+144=233\\
144+233=377\\
233+377=610\\
377+610=987\\
610+987=1597\ \ldots
\end{array}$$

The Fibonacci sequence is the oldest known (recursive) recurrent sequence. There is no evidence that Fibonacci knew of this relationship, but it is surely assumed that a man of his talents and insight knew the recursive relationship. It took 400 years following the publication of *Liber Abaci* before this relationship appeared in print.

Introducing the Fibonacci Numbers

These numbers were not identified as anything special during the time Fibonacci wrote *Liber Abaci*. As a matter of fact, the famous German mathematician and astronomer Johannes Kepler (1571–1630) mentioned these numbers in a 1611 publication[9] when he said that the ratios "as 5 is to 8, so is 8 to 13, so is 13 to 21 almost." Centuries passed and the numbers still went unnoticed. In the mid 1800s, the Fibonacci numbers began to capture the fascination of mathematicians. They took on their current name ("Fibonacci

[9] Brooke, Maxey. "Fibonacci Numbers and Their History Through 1900." *Fibonacci Quarterly*, vol. 2 (April 1964), p. 149.

numbers") from François Édouard Anatole Lucas[10] (1842–1891), a French mathematician who later devised his own sequence by following the pattern set by Fibonacci (Figure 5-4). The Lucas numbers form a sequence and similar to the Fibonacci numbers. Instead of starting with 1, 1, 2, 3, Lucas began with 1, 3, 4, 7, 11. Today, Fibonacci numbers still hold the fascination of mathematicians around the world, as they are found in part, in architecture, biology, and other forms of everyday life as we will see later in this chapter.

Figure 5-4 François Édouard Anatole Lucas

Still, one may ask, what is so special about these numbers? Before we explore the vast variety of examples in which we find the Fibonacci numbers, let us begin by inspecting this famous Fibonacci number sequence and some of its remarkable properties.

We will use the symbol F_7 to represent the 7th Fibonacci number, and F_n to represent the *n*th Fibonacci number, or the general Fibonacci number. Let us look at the first 30 Fibonacci numbers, shown in

[10] Lucas is also well known for his invention of the *Tower of Hanoi* puzzle, which he published in 1883 under the name of M. Claus (notice that Claus is an anagram of Lucas), and other mathematical recreations.

His four-volume work on recreational mathematics (1882–94) has become a classic. Lucas died as the result of a freak accident at a banquet when a plate was dropped and a piece flew up and cut his cheek. He died of erysipelas a few days later.

Figure 5-5. Notice, as we indicated earlier after the first two 1s, each number is the sum of the two preceding numbers.

$F_1 = 1$	$F_{16} = 987$
$F_2 = 1$	$F_{17} = 1597$
$F_3 = 2$	$F_{18} = 2584$
$F_4 = 3$	$F_{19} = 4181$
$F_5 = 5$	$F_{20} = 6765$
$F_6 = 8$	$F_{21} = 10946$
$F_7 = 13$	$F_{22} = 17711$
$F_8 = 21$	$F_{23} = 28657$
$F_9 = 34$	$F_{24} = 46368$
$F_{10} = 55$	$F_{25} = 75025$
$F_{11} = 89$	$F_{26} = 121393$
$F_{12} = 144$	$F_{27} = 196418$
$F_{13} = 233$	$F_{28} = 317811$
$F_{14} = 377$	$F_{29} = 514229$
$F_{15} = 610$	$F_{30} = 832040$

Figure 5-5

More on the Fibonacci Numbers can be found in *The (Fabulous) Fibonacci Numbers* by A.S. Posamentier and I. Lehmann (Amherst, NY, Prometheus Books, 2007).

One of the spectacular "coincidences" is that the Fibonacci numbers are closely related to the Golden Ratio, as we will see in the following.

Applications of the Fibonacci Numbers and the Golden Ratio

We are now about to embark on one of the most beautiful phenomenon in mathematics, the Golden Ratio, which has a remarkable relationship to the Fibonacci numbers, and which permeates more fields of mathematics than almost any other topic. As was mentioned earlier, the original problem posed by Fibonacci in his book *Liber Abaci* asks for the number of pairs of rabbits accumulating each month and leads to the sequence: l, l, 2, 3, 5, 8, 13, 21, 34, 55, 89, 144, ...

To best enchant you with the many applications of the Fibonacci numbers, you ought to have various species of pine cones, a pineapple, a sunflower, and other naturally occurring spirals.

We begin by dividing each number in the Fibonacci sequence by its successor to develop a series of fractions:

$$\frac{1}{1}, \frac{1}{2}, \frac{2}{3}, \frac{3}{5}, \frac{5}{8}, \frac{8}{13}, \frac{13}{21}, \frac{21}{34}, \frac{34}{55}, \frac{55}{89}, \frac{89}{144}, \ldots .$$

We will return to the sequence regularly in the following discussion.

Can you determine a relationship between these numbers and the leaves of a plant you might have accessible to you? From the standpoint of Fibonacci numbers, one may observe two items:

1) thc number of leaves it takes to go (rotating about the stem) from any given leaf to the next one "in a similarly direction" (that is, above it and in the same direction) on the stem; and
2) the number of revolutions as onc follows the leaves going from one leaf to another one "similarly placed." In both cases, these numbers will turn out to be the Fibonacci numbers.

In the case of leaf arrangment, the following notation is used: $\frac{3}{8}$ means that it takes three revolutions and eight leaves to arrive at the next leaf "similarly placed." In general, if we let r equal the number of revolutions, and s equal the number of leaves it takes to go from any given leaf to one "similarly placed", then $\frac{r}{s}$ will be the *phyllotaxis* (the arrangement of leaves in plants). Consider Figure 5-6, where the plant ratio is $\frac{5}{8}$.

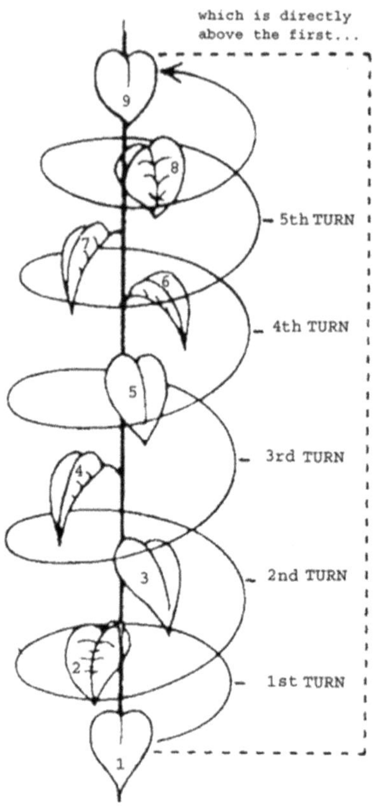

Figure 5-6

Pine cones also exhibit Fibonacci numbers. Consider the bracts on the cone to be modified leaves compressed into a smaller space. Upon observation of the cone, one can notice two spirals, one to the left (clockwise) and the other to the right (counterclockwise). One spiral increases at a sharp angle, while the other spiral increases more gradually. Consider the steep spirals and count them, as well as the spirals that increase gradually. Both numbers should be Fibonacci numbers. For example, a particular pine cone has 5 clockwise spirals and 8 counterclockwise spirals. Other pine cones may have different

Fibonacci ratios. One might wish to examine the daisy or sunflower to see where the Fibonacci ratios apply.

We notice that the ratio of consecutive Fibonacci numbers approach the Golden Ratio. The early Fibonacci ratios are:

$$\frac{2}{3} = .666667$$
$$\frac{3}{5} = .600000.$$

Then as we go further along the sequence of Fibonacci numbers the ratios begin to approach the Golden Ratio more accurately, which is represented by ϕ:

$$\phi \approx \frac{89}{144} = .618056$$
$$\phi \approx \frac{144}{233} = .618026.$$

To be more accurate, the Golden Ratio

$$\phi \approx .6180339887498948482045868343656 \ldots .$$

Geometrically, point B in Figure 5-7 divides line AC into the Golden Ratio, as

$$\frac{AB}{BC} = \frac{BC}{AC} \approx 0.618034\ldots$$

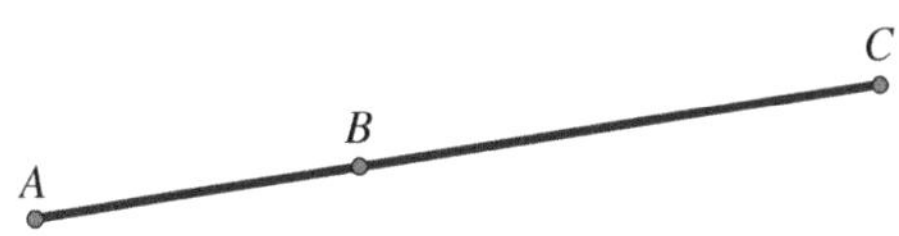

Figure 5-7

Although we will consider the *Golden Rectangle* in greater detail further on, we preview in Figure 5-8, which depicts rectangles whose dimensions are chosen so that the ratio of

$$\frac{width}{length}$$

is the Golden Ratio

$$\frac{w}{l}=\frac{l}{w+l},$$

beginning with the Golden Rectangle *ABGD*.

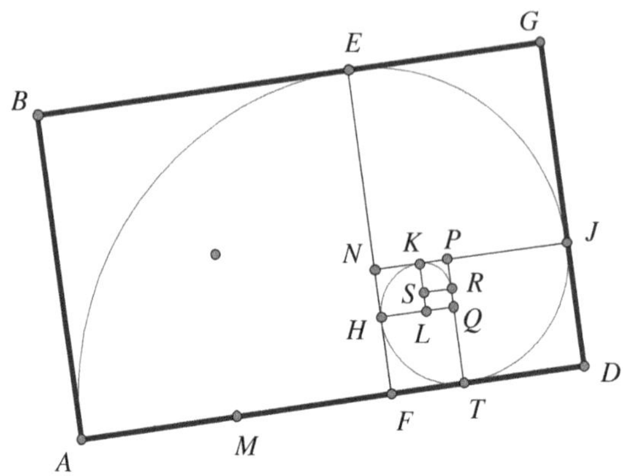

Figure 5-8

The rectangle *ABGD* in Figure 5-8 is divided by a line *EF* into a square *ABEF* and a Golden Rectangle *EFDG*. If we keep partitioning each new Golden Rectangle in the same way, we can construct a "logarithmic spiral" in the successive squares. This type of curve is frequently found in the arrangements of seeds in flowers or in the shapes of seashells and snails. You ought to find illustrations which show these spirals, such as that shown in Figure 5-9.

Figure 5-9

For another example of the Fibonacci sequence in nature, one should consider the pineapple. A pineapple has three distinct spirals of hexagons: a group of 5 spirals winding gradually in one direction, a second group of 13 spirals winding more steeply in the same direction, and a third group of 8 spirals winding in the opposite direction. Each group of spirals consists of a Fibonacci number. Each pair of spirals interacting can generate the Golden Ratio. Figure 5-10 shows a representation of the pineapple with the scales numbered in order. This order is determined by the relative distance of each hexagon from the bottom. That is, the lowest hexagon is numbered 0, the next higher one is numbered 1. Note hexagon 42 is slightly higher than hexagon 37.

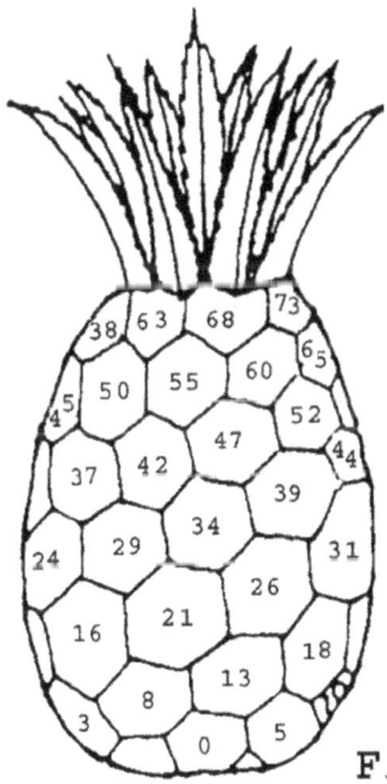

Figure 5-10

Note the three distinct sets of spirals in Figure 5-10, which cross each other, starting at the bottom. One spiral is the 0, 5, 10, etc., sequence, which increases at a slight angle. The second spiral is the 0, 13, 26, etc., sequence, which increases at a steeper angle. The third spiral has the 0, 8, 16, etc., sequence, which lies in the opposite direction from the other two. In each case, the differences between the numbers in the sequence (5 for the first spiral, 13 for the second spiral, and 8 for the third spiral) are all Fibonacci numbers. Different pineapples may have different sequences, but the Fibonacci numbers are generally present.

To move these applications of the Fibonacci numbers to another venue, consider the regeneration of male bees. Male bees hatch from unfertilized eggs; female bees hatch from fertilized eggs. The flow chart in Figure 5-11 will help you trace the regeneration of the male bees. The following pattern evolves:

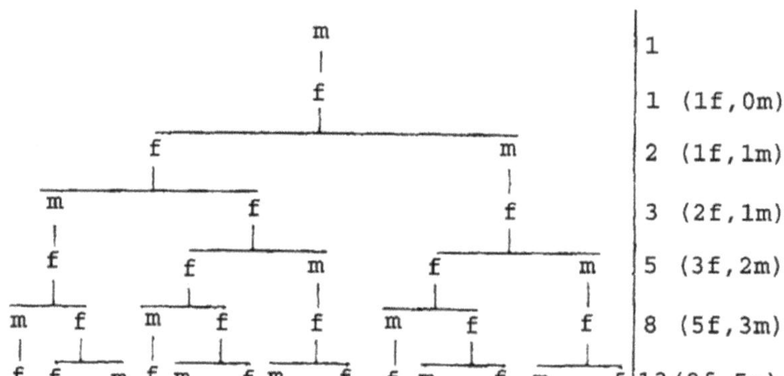

Figure 5-11

It should be obvious by now that this pattern once again features the Fibonacci sequence. As was said earlier, there are endless applications of the Fibonacci numbers, and these are but a few amazing ones.

Constructing the Golden Ratio

Reverting back to the Golden Ratio and the Golden Rectangle, recall that one ratio stands out among the rest. We partition a line segment such that we can make the following equality of two ratios, that is, a proportion, where the longer segment (L) is to the shorter segment

(S) as the entire original segment ($L + S$) is to the longer segment (L). Symbolically, this is written as

$$\frac{L}{S} = \frac{L+S}{L}.$$

Geometrically, this may be seen in him Figure 5-12. This is the *Golden Ratio*, also known as the *Golden Section* because we are "sectioning" or partitioning a line segment.

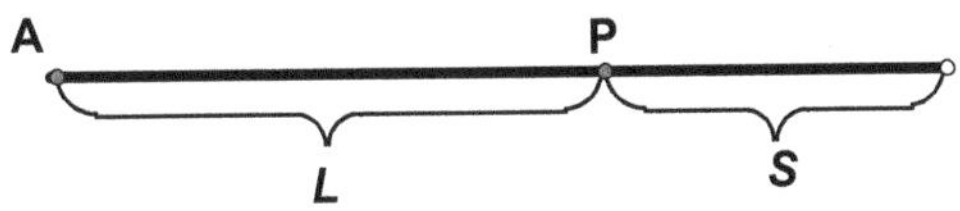

Figure 5-12

The terms Golden Ratio and Golden Section were first introduced during the nineteenth century. We believe that the Franciscan friar and mathematician Fra Luca Pacioli (ca. 1445–1517) was the first to use the term *de divina proportione* ("divine proportion") as the title of a book in 1509, while the German mathematician and astronomer Johannes Kepler (1571–1630) was the first to use the term *sectio divina* ("divine section"). Moreover, the German mathematician Martin Ohm (1792–1872) is credited for the first use of the term *Goldener Schnitt* ("golden section"). The English term *Golden Section* first appeared in an article on aesthetics by James Sully published in the 1875 edition of the Encyclopaedia Britannica.

You may be wondering what makes this ratio so outstanding that it deserves the title "Golden." This designation, which it richly deserves, will be made clear through the ensuing discussion. Let's begin by seeking to find its numerical value, which will bring us to its first unique characteristic. To determine the numerical value of the Golden Ratio $\frac{L}{S}$, we will change this equation

$$\frac{L}{S} = \frac{L+S}{L}, \text{ or } \frac{L}{S} = \frac{L}{L} + \frac{S}{L},$$

to its equivalent, where $x = \frac{L}{S}$, to get $x = 1 + \frac{1}{x}$.

We can now solve this equation for x using the quadratic formula,[11] which you may recall from high school, to obtain the numerical value of the Golden Ratio:

$$\frac{L}{S}=x=\frac{1+\sqrt{5}}{S}.$$

As we indicated earlier, the value of the Golden Ratio is commonly denoted by the Greek letter phi: ϕ.

$$\phi=\frac{L}{S}=\frac{1+\sqrt{5}}{2}\approx\frac{1+2.236067977499789696409173668731276235440}{2}$$
$$\approx 1.618033988749894848204586834365638117720.$$

Notice what happens when we take the reciprocal of $\frac{L}{S}$, namely,

$$\frac{S}{L}=\frac{1}{\phi};$$
$$\frac{1}{\phi}=\frac{S}{L}=\frac{2}{1+\sqrt{5}},$$

which when we multiply by 1 in the form of

$$\frac{1-\sqrt{5}}{1-\sqrt{5}},$$

we get:

$$\frac{2}{1+\sqrt{5}}\cdot\frac{1-\sqrt{5}}{1-\sqrt{5}}=\frac{\sqrt{5}-1}{2}\approx 0.618033988749894848204586834365638117720.$$

We then see that

$$\phi-\frac{1}{\phi}=1,$$

---

[11] The quadratic formula for solving for x in the general quadratic equation $ax^2+bx+c=0$ is $x=\dfrac{-b\pm\sqrt{b^2-4ac}}{2a}$.

which is a unique relationship of a number and its reciprocal. From the normal relationship of reciprocals, the product of ϕ and $\frac{1}{\phi}$ is also equal to 1, that is,

$$\phi \cdot \frac{1}{\phi} = 1.$$

Therefore, we have two numbers, ϕ and $\frac{1}{\phi}$, whose difference and product are equal to 1. These are the only two numbers for which this is true! By the way, you might have noticed that their sum

$$\phi + \frac{1}{\phi} = \sqrt{5},$$

since

$$\frac{\sqrt{5}+1}{2} + \frac{\sqrt{5}-1}{2} = \sqrt{5}.$$

Having defined the Golden Ratio numerically, we shall now *construct* it geometrically. There are several ways to construct the Golden Section of a line segment. To avoid confusion, we will use the term Golden Ratio to refer to the numerical value of ϕ, and the term Golden Section to refer to the geometric division of a segment into the ratio ϕ.

Golden Section Construction 1

One popular procedure to construct a Golden Section is to begin with a unit square *ABCD*, shown in Figure 5-13, with midpoint *M* on side *AB*. Then draw a circular arc with center *M* and radius *MC* cutting the

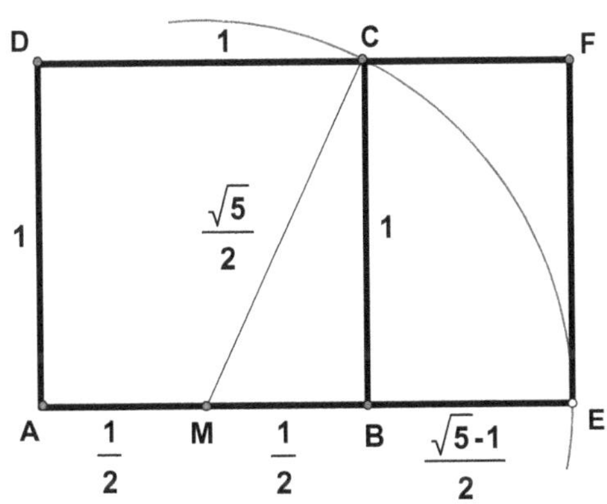

Figure 5-13

extension of side AB at point E. We now can claim that the line segment AE is partitioned into the Golden Section at point B. This, of course, will be substantiated.

To verify this claim, we would have to apply the definition of the Golden Section

$$\frac{AB}{BE}=\frac{AE}{AB},$$

and see if it, in fact, holds true. By substituting the values obtained by applying the Pythagorean theorem to ΔMBC as shown in Figure 5-13, we get the following:

$$MC^2=MB^2+BC^2=\left(\frac{1}{2}\right)^2+1^2=\frac{1}{4}+1=\frac{5}{4};$$

therefore,

$$MC=\frac{\sqrt{5}}{2}.$$

It follows that

$$BE=ME-MB=MC-MB=\frac{\sqrt{5}}{2}-\frac{1}{2}=\frac{\sqrt{5}-1}{2},$$

and

$$AE=AB+BE=1+\frac{\sqrt{5}-1}{2}=\frac{2}{2}+\frac{\sqrt{5}-1}{2}=\frac{\sqrt{5}+1}{2}.$$

We then can find the value of $\frac{AB}{BE}=\frac{AE}{AB}$, that is

$$\frac{1}{\frac{\sqrt{5}-1}{2}}=\frac{\frac{\sqrt{5}+1}{2}}{1},$$

which turns out to be a true proportion, since the cross products are equal. That is,

$$\left(\frac{\sqrt{5}-1}{2}\right)\cdot\left(\frac{\sqrt{5}+1}{2}\right)=1\cdot 1=1.$$

We can also see from Figure 5-13 that point B divides the line segment AE into an inner Golden Section, since

$$\frac{AB}{AE}=\frac{1}{\frac{\sqrt{5}-1}{2}}=\frac{\sqrt{5}+1}{2}=\phi.$$

Meanwhile, point E can be said to divide the line segment AB into an outer Golden Section, since

$$\frac{AE}{AB}=\frac{1+\frac{\sqrt{5}-1}{2}}{1}=\frac{\sqrt{5}+1}{2}=\phi.$$

You ought to take notice of the shape of the rectangle $AEFD$ in Figure 5-13. The ratio of the length to the width is the Golden Ratio:

$$\frac{AE}{EF}=\frac{\frac{\sqrt{5}+1}{2}}{1}=\frac{\sqrt{5}+1}{2}=\phi.$$

This appealing shape is called the Golden Rectangle, which was alluded to earlier.

For centuries, artists and architects have identified the perfectly shaped rectangle, the rectangle most pleasing to the eye, as the Golden Rectangle. The desirability of this rectangle has been borne out by numerous psychological experiments. For example, Gustav Fechner (1801–1887), a German experimental psychologist inspired by Adolf Zeising's book *Der Goldene Schnitt*,[12] investigated the psychological aesthetic appeal of the Golden Rectangle, and published his findings in 1876.[13] Fechner made thousands of measurements of commonly seen rectangles, such as playing cards, writing pads, books, windows, etc. He found that most had a ratio of length to width that was close to ϕ. He also tested people's preferences, and found that most people preferred the shape of the Golden Rectangle.

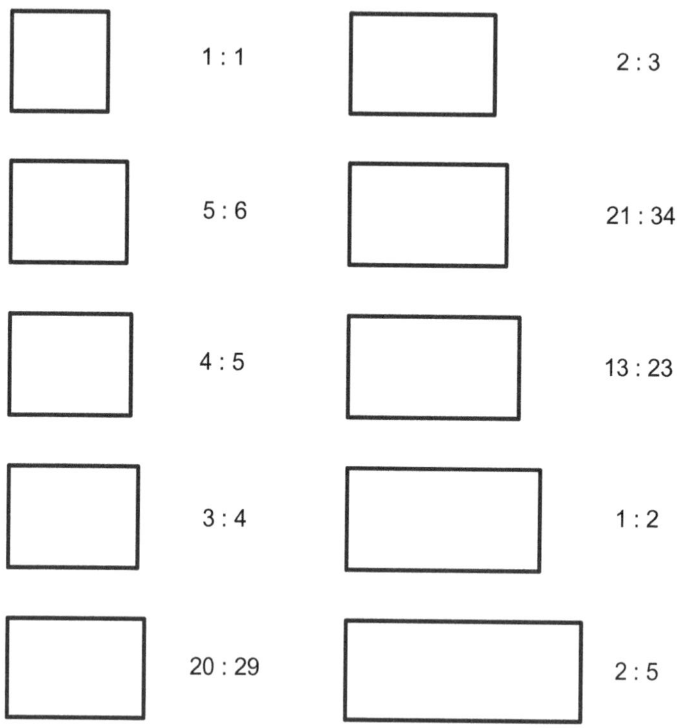

Figure 5-14 Fechner's Rectangle

Fechner asked 228 men and 119 women which of the rectangles shown in Figure 5-14 is the most aesthetically pleasing. Which

---

[12] *Der Goldene Schnitt* (The Golden Ratio) was published posthumously by the Leopoldinisch-Carolinische Akademie, Halle, Germany, 1884.

[13] Fechner, Gustav Theodor. *Zur Experimentalen Ästhetik* (On Experimental Aesthetics). Leipzig, Germany: Breitkopf & Haertel, 1876.

rectangle would you choose? Rectangle 1:1 is too much like a square, and not considered by the general public to be representative of a "rectangle." It is, after all, a square! On the other hand, rectangle 2:5 (the other extreme) is uncomfortable to look at since it requires the eye to scan it horizontally. The shape of rectangle 21:34 can be appreciated at a single glance. Fechner's findings seem to bear this out. The results that Fechner reported are shown in Figure 5-15.

Ratio of sides of rectangle	Percent response for best rectangle	Percent response for worst rectangle
1:1 = 1.00000	3.0	27.8
5:6 = .83333	.02	19.7
4:5 = .80000	2.0	9.4
3:4 = .75000	2.5	2.5
20:29 = .68966	7.7	1.2
2:3 = .66667	20.6	0.4
21:34 = .61765	**35.0**	**0.0**
13:23 = .56522	20.0	0.8
1:2 – .50000	7.5	2.5
2:5 = .40000	1.5	35.7
	100.00	100.00

Figure 5-15

Fechner's experiment has been repeated with variations in methodology many times, further supporting his results. For example, in 1917 Edward Lee Thorndike (1874–1949), the American psychologist and educator, carried out similar experiments, with analogous results. In general, the rectangle with the ratio of 21:34

was most preferred. Do those numbers look familiar? Yes, once again the Fibonacci numbers appear. The ratio

$$\frac{21}{34} = 0.6\overline{1764705882352941}$$

approaches the value of ϕ, and gives us the Golden Rectangle.

Displaying a Watch

We are often unaware of the many places that we experience aspects of the Golden Rectangle. In almost all advertisements for watches and clocks, the hands of the clock show approximately 10:10, as in Figure 5-16.

Figure 5-16

The hands make an angle of approximately $19\frac{1}{6}$ minutes, which is the equivalent of 115°. A quick aside to explain how we got this angle. Since 10 minutes is $\frac{1}{6}$ of an hour, the hour hand moved $\frac{1}{6}$ of the distance from "10" to "11" between 10:00 and 10:10. Therefore, the hour hand moved $\frac{1}{6}$ of 5 minute markers, or $\frac{5}{6}$ of a minute marker, making its distance from "12" $9\frac{1}{6}$ minutes. The minute hand's distance from "12" is 10 minutes. Thus, the angle at 10:10 is $19\frac{1}{6}$ minute markers.

To find the degree measure of that angle, we simply find the part of 60-minute markers that $19\frac{1}{6}$ is:

$$\frac{19\frac{1}{6}}{60}=\frac{\frac{115}{6}}{60}=\frac{115}{360}=115°.$$

Now make the points indicated by the hands at 10:10 (that is, at "2" and "$10\frac{1}{6}$") the two adjacent vertices of a rectangle with diagonals intersecting at the center of the watch face, as shown in Figure 5-17. This rectangle is very close to the Golden Rectangle, whose diagonals meet at an angle of about 116.6°, as can be seen in Figure 5-18.

Figure 5-17

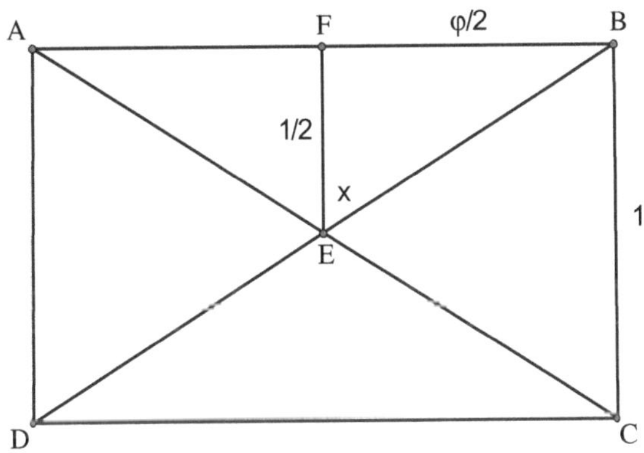

Figure 5-18

This can be determined by using the Golden Rectangle, shown in Figure 5-18, and finding the measure of the angle whose tangent function is tan

$$x=\frac{FB}{FE}=\frac{\frac{\phi}{2}}{\frac{1}{2}}=\phi.$$

That angle, x, measures 58.28°. Therefore, $\angle AEB=116.56°$, which is very close to 115°, the ideal placement of the hands of a watch. Some might have suspected that the placement of the hands at 10:10

allows the observer to see the brand name of the watch without obstruction, and has nothing to do with the Golden Rectangle. Not necessarily so. Take for another example a United States postage stamp, shown in Figure 5-19, that exhibits the "American clock."

Figure 5-19

Here, the time shown is about 8:21½, where the hands form an angle[14] of about 112.79167°. This is not exactly the angle of the diagonals of an ideal Golden Rectangle, but it is close enough for the naked eye to appreciate. The Golden Rectangle seems to have again determined the ideal appearance!

Applications of the Golden Ratio in Art

The famous Italian painter, sculptor, architect, natural scientist, and engineer Leonardo da Vinci (1452–1519) illustrated the book *De Divina Proporzione*, written by the Franciscan monk Fra Luca Pacioli[15] (ca. 1445–1517). He drew an anatomical study of the

[14] 60 min = 360°, 1 min = 6°, 10 min = 60°. Therefore, $\angle(12; 8{:}21\frac{1}{2}) = 21\frac{1}{2} \times 6° = 129°$, and $\angle(6; 21\frac{1}{2}) = 180° - 129° = 51°$. The hour hand's location: 12 h = 360°, 1 h = 30°, 10 min = $\frac{1}{6}$h = 5°, 1 min = $\frac{1}{60}$h = $\frac{1}{12}$°, $\angle(12; 8{:}00) = 8 \times 30° = 240°$, $\angle(8{:}00; 8{:}21\frac{1}{2}) = 21{,}5 \times \frac{1}{12}° = 1.79167°$. Therefore, $\angle(12; 8{:}21\frac{1}{2}) = 241.79167°$, and $\angle(6; 8{:}21\frac{1}{2}) = 241.79167° - 180° = 61.79167°$. The sum = 51° + 61.79167° = **112.79167°**.

[15] Pacioli, Luca. *De Divina Proportione*. Venezia, 1509; reprinted in Milan in 1896, Gardner Pelican, 1961.

Vitruvian man (Figure 5-20) to support Pacioli's discussion of the Roman architect Vitruvius[16] (ca. 84 BCE–27 BCE), who posited that the proportions of the human body should be the basis for architecture. The ratio of square side to circle radius in this famous picture corresponds to the Golden Ratio with a deviation of 1.7%.

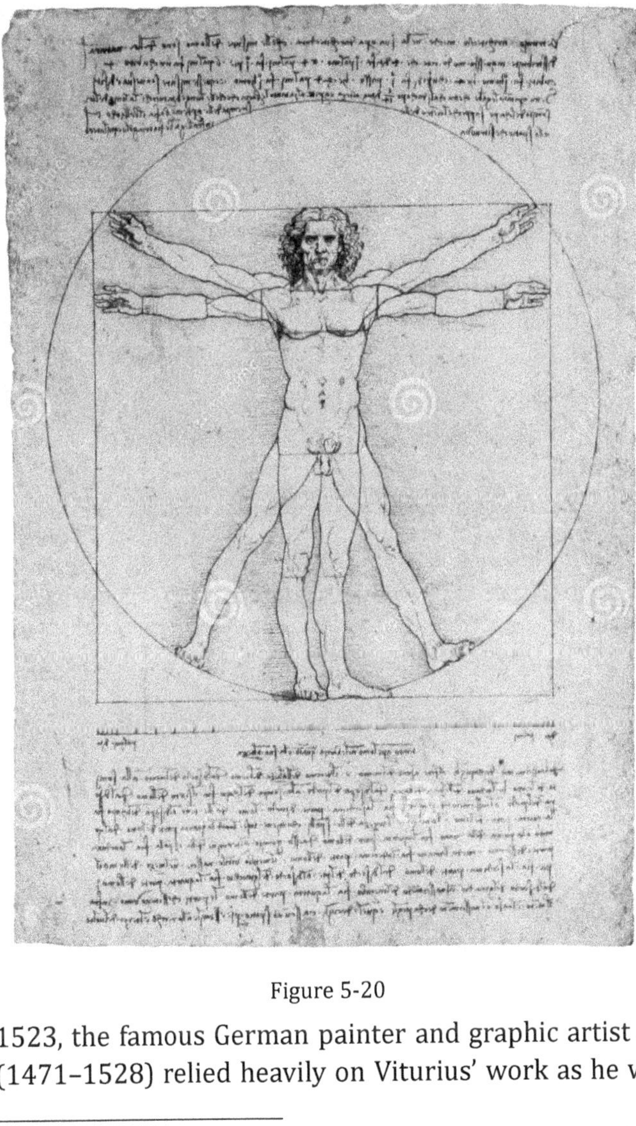

Figure 5-20

In 1523, the famous German painter and graphic artist Albrecht Dürer (1471–1528) relied heavily on Viturius' work as he wrote his

[16] Vitruvius Pollio, a Roman army technician and engineer, published ten books about architecture and civil engineering and developed a theory of proportions.

Four Books on Human Proportions. Dürer refined the earlier works of Vitruvius and Pacioli, and devised a system of proportions where the unit of measure is the human body with the parts expressed as fractions thereof.

In his books on geometry, fortress building, and human proportions, Dürer supported all of his ideas with theoretical rather than geometric arguments. As a matter of fact, his works in descriptive geometry influenced some of the greatest thinkers of the Renaissance, such as Johannes Kepler (1571–1630) and Galileo Galilei (1564–1642).

In his famous self-portrait (ca. 1500), Dürer drew himself with a head of wavy hair whose outline creates an equilateral triangle. This can be seen in Figure 5-21, where Dürer superimposed the triangle and several other guidelines over the self-portrait. The base of the equilateral triangle divides the height of the entire picture into the Golden Ratio. The chin also divides the height of the entire picture into the Golden Ratio.

By the way, you will notice in Figure 5-22 that Dürer made his self-portrait so that his head could be encased perfectly into a Golden Rectangle. This is further evidence that Dürer was aware of some of the techniques that da Vinci championed in his own artwork: perspectivity and the Golden Ratio.

The famous British art critic Sir Theodore Cook (1867–1928) analyzed the Italian Renaissance painter Sandro Botticelli's (1445–1510) painting *The Birth of Venus* (1477, Figure 5-23). The

Figure 5-21

Figure 5-22

Figure 5-23

rectangular dimensions of the picture are 278.5 cm by 172.5 cm, which yields

$$\frac{278.5}{172.5} = 1.6145,$$

which is very close to the Golden Ratio ($\phi \approx 1.618$). Also, the ratio of the width of water to the width of the sky is 1.612 cm, which is also very close to the Golden Ratio. Furthermore, the ratio between the distance from Venus's navel to heel and the distance from navel to head is approximately 1.599 cm, again very close to the Golden ratio. In many cases, we do not know if the Golden Ratio was used intentionally or serendipitously. In the 19th and 20th centuries, many scholars searched for applications of the Golden Ratio.

The question of whether architects and artists consciously used the Golden Ratio in their work was discussed by Dan Pedoe (1910–1998) in his book *Geometry and the Visual Arts.* He argues that there is hardly any proof that the Golden Ratio was used in Renaissance architecture, although it motivated great interest in the field. As they say, beauty is in the eye of the beholder, and so the beholder must make their own judgment about the use of the Golden Ratio. We will concentrate on a few paintings where the Golden Ratio appears, but there are countless additional examples that you can find in art and architecture over the past several centuries. Let us begin with one of the most famous paintings in Western civilization, Leonardo da Vinci's *Mona Lisa*, which was painted 1503–1506 and is on exhibit in the Louvre in Paris. (See Figure 5-24.)

To begin our inspection of the painting, we draw a rectangle around Mona Lisa's face and find that it is a Golden Rectangle, as shown in Figure 5-25. Furthermore, in Figure 5-26, you will notice specific points on the body of the Mona Lisa determining Golden Ratios. Since da Vinci illustrated Pacioli's book *Da Divina Proportione*, which thoroughly discussed the Golden Ratio, it can be assumed that he was consciously guided by this magnificent ratio.

Figure 5-24

Figure 5-25

Figure 5-26

The Golden Ratio in Architecture

The Parthenon in Athens, Greece, has a total height of about 60 feet, and a width of about 101 feet. Therefore, the ratio of the building's height and width is $\frac{101}{60} \approx 1.683$. We should again notice that this number is quite close to ϕ. Figure 5-27 shows that the façade, with its figuratively reconstructed pediment (dashed line), fits almost perfectly into a Golden Rectangle (bold outline).

Figure 5-27 Parthenon's Façade in Golden Rectangle

The outer rectangle of the façade, including the steps, measures 111.31 feet wide and 65.162 feet high, so it has a ratio of

$$\frac{111.31}{65.162} \approx 1.7082.$$

A rectangle with this width:height ratio can be exactly divided into a square and three additional rectangles with the ratio $\sqrt{5}-2$, which is the same as $2\phi-1$, as shown in Figure 5-28. The Golden Ratio occurs multiple times in the Parthenon's façade. The rectangle shown in Figure 5-28 has measurements of a and

$$a+3a(2\phi-1)=a(6\phi-2).$$

The ratio of its width to its height can be calculated as

$$\frac{a(6\phi-2)}{a}=6\phi-2\approx 1.7082.$$

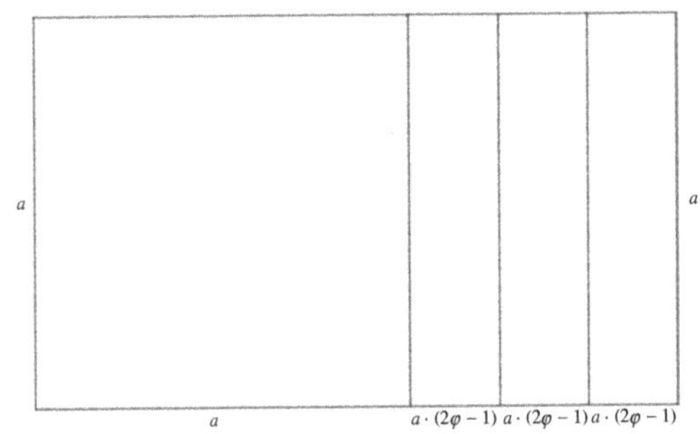

Figure 5-28 The Number φ in the Parthenon's Façade

Let A and B be the endpoints of the horizontal baseline of the ground floor. The points D and C are perpendicularly above the points A and B, respectively, at the lower edge of the gable, thereby creating the rectangle $ABCD$, as shown in Figure 5-29. This rectangle consists of four Golden Rectangles placed side-by-side, so that rectangle $ABCD$ has a side-ratio of 4ϕ.

Figure 5-29 Golden Rectangles in Parthenon's Façade 1

Now scale down *ABCD* to a similar rectangle *EFGH*, shown in Figure 5-30, where *EF* is still the horizontal baseline of the ground floor and *GH* is defined by the upper edge of the columns. Due to its similarity to *ABCD* (side-ratio $= 4\phi$), it can also be subdivided into four Golden Rectangles placed side-by-side.

Figure 5-30 Golden Rectangles in Parthenon's Façade 2

This newly generated rectangle helps us understand the position of the columns, which were not placed randomly, but follow strict geometrical requirements related to ϕ.

Let's consider one of the four Golden Rectangles and partition it into 3 congruent smaller rectangles—which are, of course, all Golden Rectangles as well. We can subdivide *EFGH* into 12 congruent Golden Rectangles. Figure 5-31 shows the 11 vertical separation lines. It is now discernible that the inner columns are always spaced exactly two lines apart.

Suppose we let the height of a column $EH = h$. Then $EF = 4\phi h$. The distance between two columns is one-sixth of this length, that is, $\frac{2}{3}\phi h$. That means that the distance between two columns is $\frac{2}{3}\phi \approx 0.412$ times their height. But the façade of the Parthenon can be further subdivided, revealing geometrical peculiarities related not to ϕ, but

Figure 5-31 Golden Rectangles in Parthenon's Façade 3

to $\sqrt{5}$. The presence of $\sqrt{5}$ is illustrated by Figure 5-32, where the numbers inside the rectangles indicate their width:height ratio. The bottom line of the *epistyle* (the long beam lying on top of the columns) divides the rectangle into an upper part, consisting of the epistyle and the gable, and a lower part, consisting of the columns. The whole

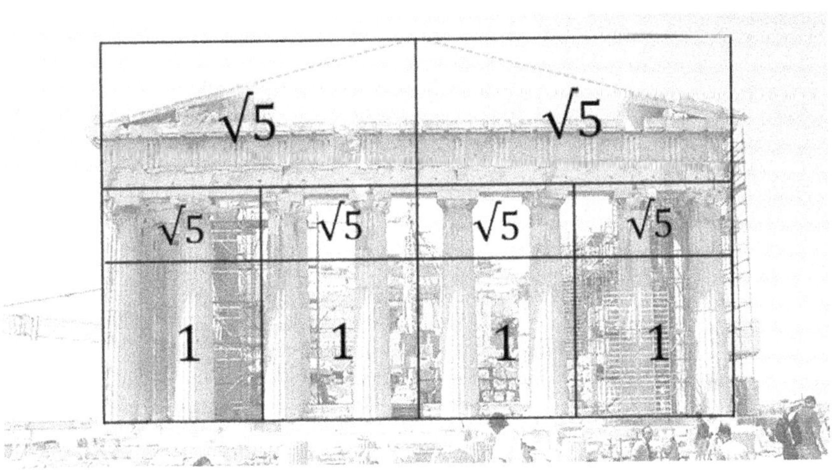

Figure 5-32 Subdivision of the Pantheon's Façade

upper part is composed of two identical rectangles with ratio $\sqrt{5}$. Only the small outmost sides of the gable on the left and right are outside of these rectangles. The lower part of the rectangle can be divided vertically into four smaller congruent rectangles, each of which has the proportion of

$$\frac{1}{\sqrt{5}}+1,$$

so every single one of these four rectangles can be cut into a square beneath a rectangle with ratio $\sqrt{5}$.

Having investigated numerous aspects of Fibonacci numbers, the Golden Ratio, and their appearances in nature, art, and geometry, you should have a greater appreciation for mathematics. Keep searching for other applications of mathematics, the most important subject in our culture.

Appendix

The Justification for the Varignon Parallelogram

To prove that this inner quadrilateral formed by joining the midpoints successively of the larger parallelogram, is actually a parallelogram, we rely on the basic theorem that the lines joining the midpoints of two sides of a triangle are parallel to the third side and half its length. In Figure A-1, for triangle ABD we have $KN \| BD$ and $KN = \frac{1}{2}BD$, and for triangle BCD we have $LM \| BD$ and $LM = \frac{1}{2}BD$, so that $KN \| LM$ and $KN = LM$. Therefore, $KLMN$ is a parallelogram since one pair of opposite sides are equal and parallel.

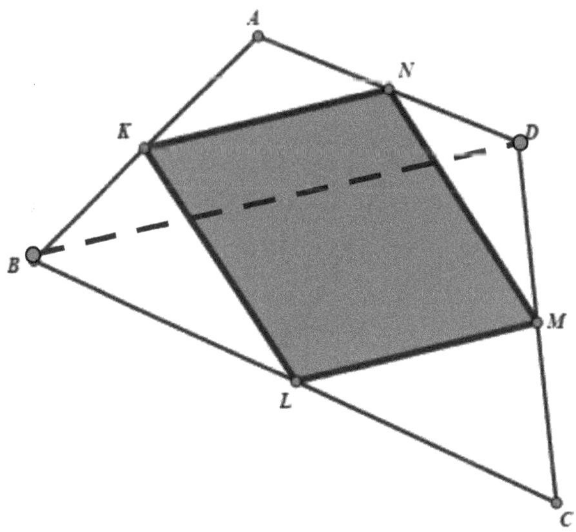

Figure A-1

The Justification for the Varignon Parallelogram's Special Properties

The sum of the two lines joining the midpoints of adjacent sides of a quadrilateral is equal to the length of the diagonal parallel to them. In Figure A-2, we have $KN + LM = BD$. Similarly, we have $KL + MN = AC$, so we can conclude that the perimeter of the parallelogram $KLMN$ is equal to the sum of the diagonals of the original quadrilateral. Symbolically, $KN + LM + KL + MN = BD + AC$.

We can also show that the Varignon parallelogram is one-half the area of the original quadrilateral. We once again refer to Figure A-2, where $\Delta AKN \sim \Delta ABD$ and where the $area\Delta AKN = \frac{1}{4}area\Delta ABD$. Similarly, we have $\Delta LCM \sim \Delta BCD$ and $area\Delta LCM = \frac{1}{4}area\Delta BCD$. Therefore, $area\Delta AKN + area\Delta LCM = \frac{1}{4}area\Delta ABD + \frac{1}{4}area\Delta BCD = \frac{1}{4}areaABCD$. When we repeat this procedure using diagonal AC, we find that $area\Delta KBL + area\Delta NDM = \frac{1}{4}area\Delta ABC + \frac{1}{4}area\Delta ACD = \frac{1}{4}areaABCD$. Therefore, $area\Delta AKN + area\Delta LCM + area\Delta KBL + area\Delta NDM = \frac{1}{2}areaABCD$, which is to say that the parallelogram is one-half the area of the original quadrilateral.

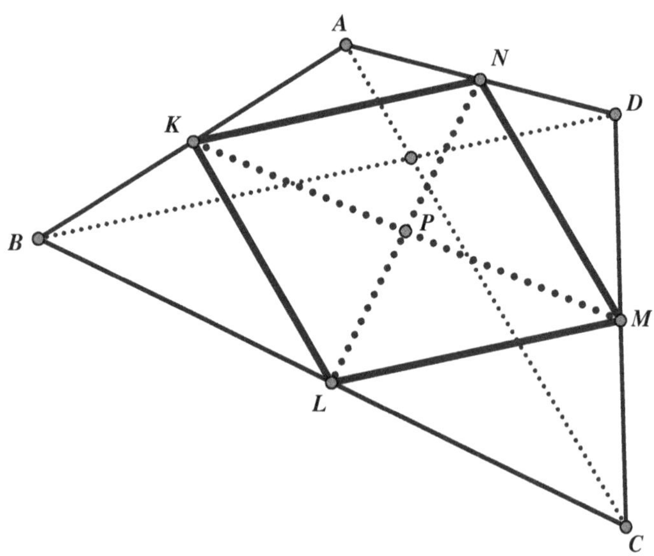

Figure A-2

This also provides another feature of a quadrilateral, namely, that the lines joining the midpoints of the opposite sides bisect each other, as they are also the diagonals of the Varignon parallelogram. This determines the common midpoint *P* of *KM* and *LN*, as shown in Figure A-2.

A Justification for an Alternative Way of Creating a Varignon Parallelogram

The length of a line joining the midpoints of two sides of triangle is equal to one-half the length of the third side and parallel to the third side. Consider triangles *ABC* and *DBC* in Figure A-3, where $RQ \parallel BC$ and $PS \parallel BC$. Therefore, $RQ \parallel PS$. Also, $RQ = \frac{1}{2}BC$ and $PS = \frac{1}{2}BC$. Therefore, $RQ = PS$. Thus, quadrilateral *PQRS* is a parallelogram since one pair of sides are equal and parallel.

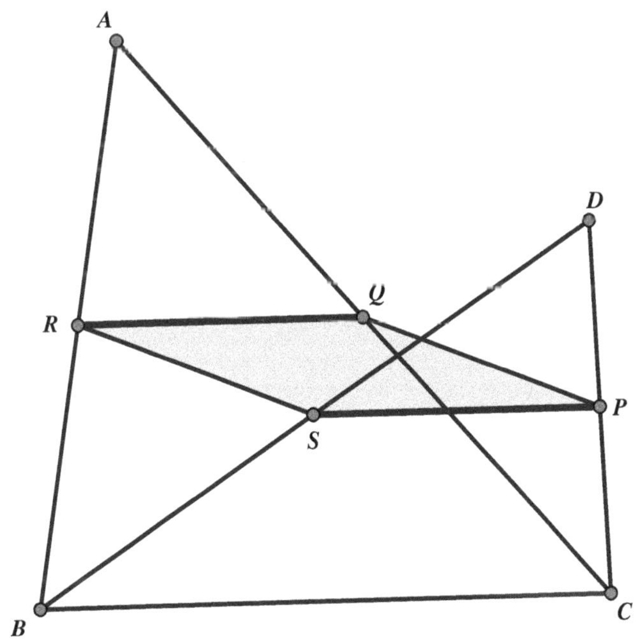

Figure A-3

A Justification for a Random Point in a Parallelogram Determinining an Equal Area Triangle Relationship

Perhaps the simplest way to justify this challenge is to consider the parallel lines shown in Figure A-4, with the area of each small triangular region marked with m, n, p, and q. Through point P two lines are drawn parallel to the sides of the parallelogram, partitioning $ABCD$ into four parallelograms. Each of these smaller parallelograms is divided in half by the four lines emanating from point P. $Area\Delta APB + Area\Delta DPC = (n+q)+(m+p) = m+n+p+q$, and similarly, $Area\Delta BPC + Area\Delta APD = (p+q)+(m+n) = m+n+p+q$. Therefore, $Area\Delta APB + Area\Delta DPC = Area\Delta BPC + Area\Delta APD$.

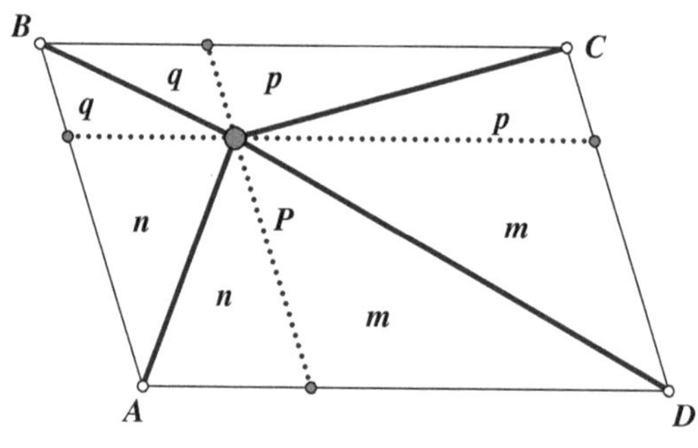

Figure A-4

A Clever Solution for Parallelogram Comparisons

A common approach attempts to solve the problem by looking for congruent triangles with equal areas. This method will lead nowhere. A clever method, although rather "off the beaten path," is to draw the line segment PD, shown in Figure A-5.

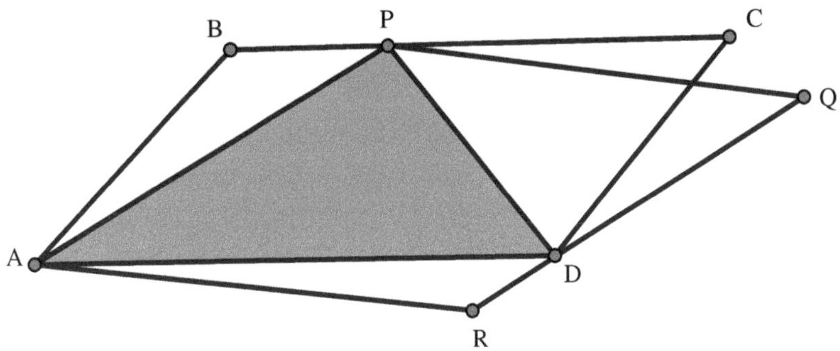

Figure A-5

Then notice that triangle *APD* is one-half the area of each of the two parallelograms, since in each case it shares a base and a height with each of the parallelograms. That is, triangle *APD* shares a base *AD* and the altitude from *P* to *AD* with parallelogram *ABCD*, which makes triangle *APD* half the area of the parallelogram. Similarly, triangle *APD* shares a common base *AP* and the altitude from *D* to *AP* with parallelogram *APQR*; therefore, triangle *APD* is also one-half the area of parallelogram *APQR*. The two parallelograms are thus equal in area, as they are each double the area of triangle *APD*.

Although this is a clever approach to a rather challenging problem, there is an even more elegant solution. Recall that when the problem was posed, we were merely told that point *P* was on side *BC*, but not where along the side it was placed. We can consider an extreme case in which *P* overlaps point *B*. Similarly, point *D*, which was to be placed on side *RQ*, could just as easily have been placed to overlap point *R*. Under these circumstances, which certainly satisfy the original problem's constraints, the two parallelograms would overlap and have the same area. Therefore, the area of parallelogram *APQR* is 18.

The Justification of the Unexpected Triangle That Is Half the Area of a Trapezoid

We begin by drawing a line through the midpoint P of DA, parallel to base AB, to intersect BC at point R. We also draw CH perpendicular to AB at point H, and intersecting PR at point G. For convenience, the lengths of the line segments are marked as shown in Figure A-6: $AB = a$, $CD = c$, $CG = h = GH$, and $PR = m$. Since area $\Delta CPR = \frac{1}{2}mh =$ area ΔBRP, the sum of the areas of the two triangles ΔCPR and ΔBRP is area $\Delta BCP =$ area $\Delta CPR +$ area $\Delta BRP = mh$. However, since P is the midpoint of DA, we have area $ABCD = 2mh$. The sum of the areas of the two triangles is thus half the area of the trapezoid, with area $\Delta BCP = \frac{1}{2}$area $ABCD$.

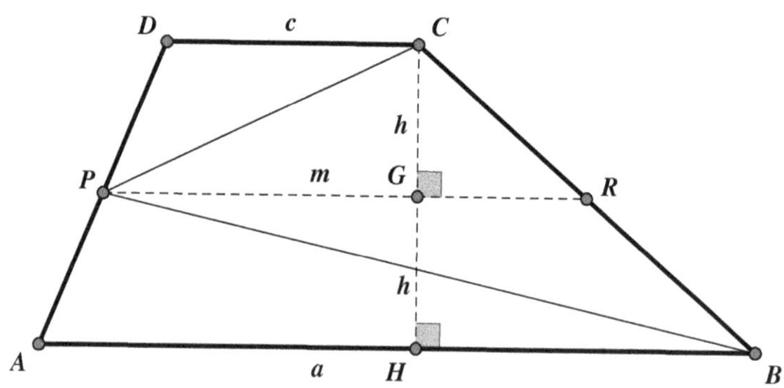

Figure A-6

Justifying a Truly Unexpected Collinearity

In Figure A-7, we begin by noticing that the two equilateral triangles BEC and DFC have the same length base so that $EC = FC$.

Since $\angle FCE = \angle FCD + \angle DCB - \angle ECB = 60° + 90° - 60° = 90°$, we also note that $\angle CEF = \frac{1}{2}(180° - \angle FCE) = \frac{1}{2}(180° - 90°) = 45°$, which determines that triangle FEC is an isosceles right triangle. Also, triangle ABE is an isosceles triangle with $\angle EBA = 30°$, and therefore, $\angle AEB = \frac{1}{2}(180° - \angle EBA) = \frac{1}{2}(180° - 30°) = 75°$. We know that $\angle BEC = 60°$. Therefore, $\angle AEB + \angle BEC + \angle CEF = 75° + 60° + 45° = 180°$, and A, E, and F are thus, collinear.

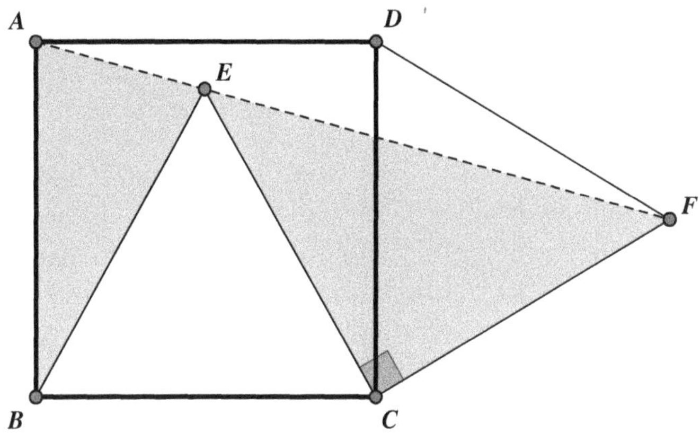

Figure A-7

A Justification That Two Equal Circles Produce an Isosceles Triangle in a Rather Strange Way

In Figure A-8, we draw the common chord *AB* of the two circles as well as the radii *OA*, *OB*, *QA*, and *QB*. Since triangles *OBA* and *QAB* share a side *AB*, and their other sides are of equal length, we have $\Delta OBA \cong \Delta QAB$, and thus $\angle BOA = \angle AQB$. In circle *O* we have $\angle BXA = \frac{1}{2}\angle BOA$, and in circle *Q* we have $\angle AYB = \frac{1}{2}\angle AQB$. Therefore, $\angle BXY = \angle BXA = \frac{1}{2}\angle BOA = \frac{1}{2}\angle AQB = \angle AYB = \angle XYB$, and with $\angle BXY = \angle XYB$, we see that ΔBYX is isosceles.

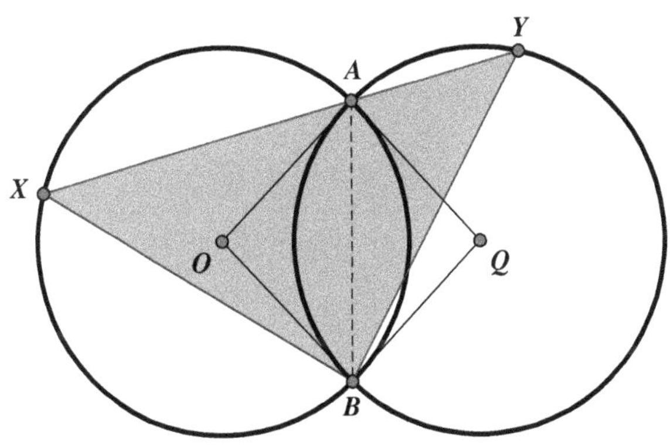

Figure A-8

A Justification for Having Two Equal Circles Generate Another Isosceles Triangle

In Figure A-9, we have drawn the common chord AB of the two circles and radii OA, OB, QA, and QB. Since triangles ΔOBA and ΔQAB share a side AB, and their other sides are of equal length, we have $\Delta OBA \cong \Delta QAB$ and $\angle BOA = \angle AQB$. In circle O, we have $\angle AXB = 180° - \frac{1}{2}\angle BOA$ and $\angle BXY = 180° - \angle AXB = \frac{1}{2}\angle BOA$. In circle Q, we have $\angle AYB = \frac{1}{2}\angle AQB$, and this gives us $\angle BXY = \frac{1}{2}\cdot\angle BOA = \frac{1}{2}\cdot\angle AQB = \angle AYB = \angle XYB$. With $\angle BXY = \angle XYB$, we see that ΔBYX is isosceles.

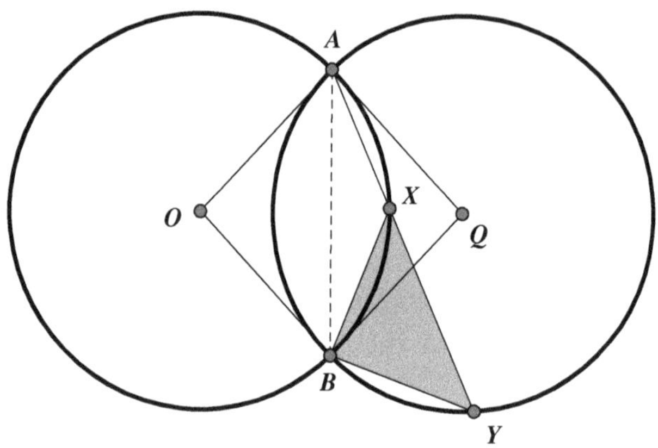

Figure A-9

A Justification of the Triangles with a Common Base Producing Concurrent Angle Bisectors

Since the angles at points C, D, E, and F are of equal measure, they are concyclic on the common chord AB of the circumcircle of triangle ABC, as shown in Figure A-10. Each of these angles measures one-half of the measure of arc *AB*. The bisectors of each of these angles will also meet at the midpoint *P* of arc *AB*, and, therefore, the bisectors are concurrent.

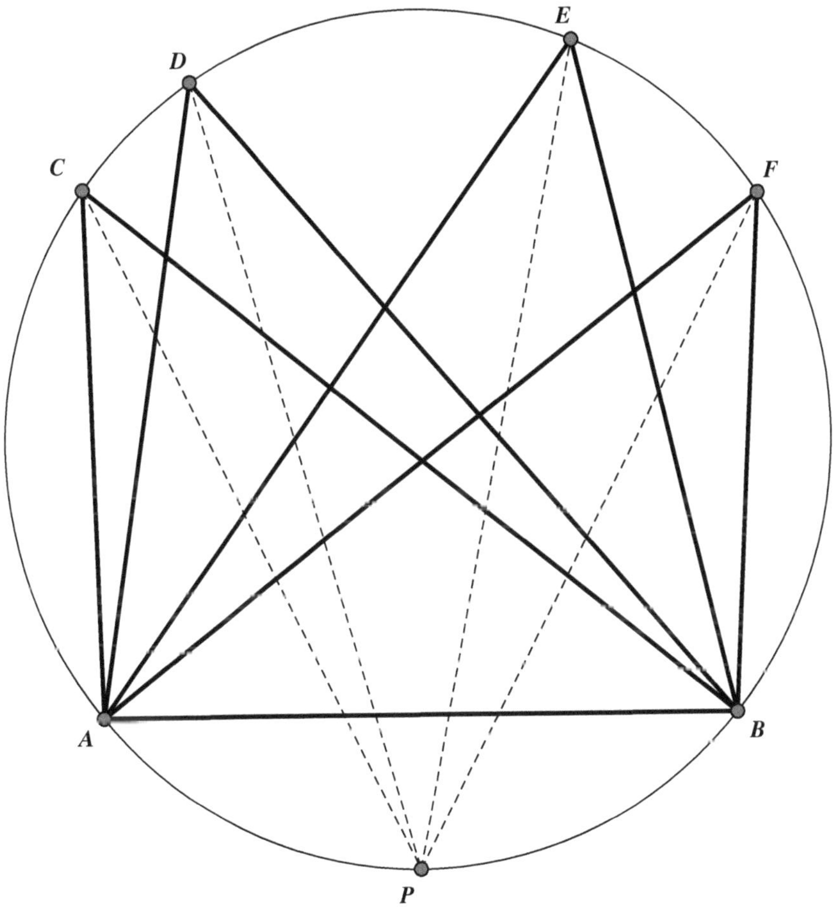

Figure A-10

A Justification for Creating Unexpected Supplementary Angles

In Figure A-11, we first draw chords *AD*, *DB*, and *BC*. On chord *AD*, we have $\angle AOD = 2\angle ABD$, since both angles are measured by arc *AD*. Similarly, on chord *BC*, we have $\angle BOC = 2\angle BDC$. Also, in right triangle *PDB*, we have $\angle PBD = 90° - \angle BDP$. In summary, we obtain

$$\angle AOD = 2\cdot \angle ABD = 2\cdot \angle PBD = 2\cdot(90° - \angle BDP) = 180° - 2\cdot \angle BDC$$
$$= 180° - 2\cdot\frac{1}{2}\cdot \angle BOC = 180° - \angle BOC.$$

We see those angles $\angle AOD$ and $\angle BOC$ are supplementary, as we set out to prove.

This is the way one typically solves the problem. However, a more elegant solution would be to realize that when two chords intersect, the angle formed is half the sum of the intercepted arcs. Therefore, since $\angle BPC = 90°$, the sum of $\overset{\frown}{AD} + \overset{\frown}{BC} = 180°$. However, the central angles $\angle AOD = \overset{\frown}{AD}$, and $\angle BOC = \overset{\frown}{BC}$. Thus, we have $\angle AOD + \angle BOC = 180°$.

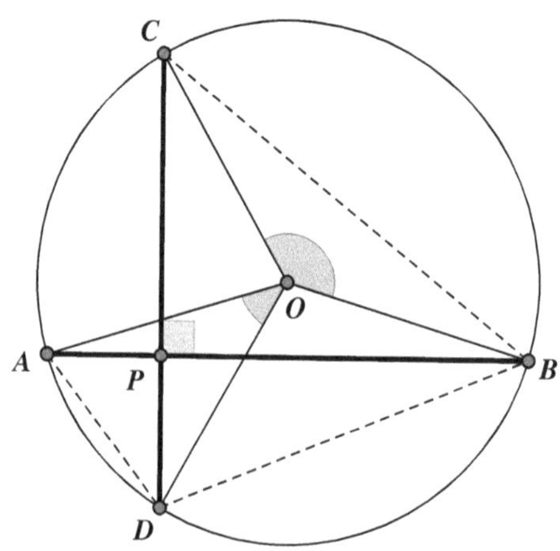

Figure A-11

Justifying the Circumscribable Quadrilateral

To simplify the first part of this proof, we have labeled the tangents as shown in Figure A-12, where quadrilateral $ABCD$ has an inscribed circle with center O. Since tangents drawn to the same circle from the same external point are equal, we have $AH = AE = w$, $BE = BF = x$, $CF = CG = y$, and $DG = DH = z$. Therefore, we have $AB + CD = (w + x) + (y + z) = (x + y) + (z + w) = BC + DA$, and thus, $AB + CD = BC + DA$.

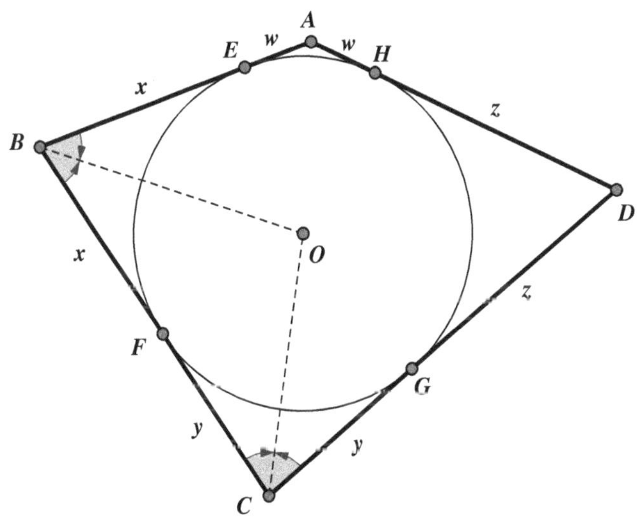

Figure A-12

In order to prove the converse, we assume that $ABCD$ is a quadrilateral in which we have $AB + CD = BC + DA$. We wish to show that there must exist a circle tangent to all four sides of the quadrilateral. We now consider Figures A-13 and A-14, where, without loss of generality, we assume that DA is the shortest of the sides of quadrilateral $ABCD$. We then have either $AB = DA$ or $AB > DA$. If $AB = DA$, we also have $BC = AB + CD - DA = AB + CD - AB = CD$, and quadrilateral $ABCD$

is therefore a deltoid (or a kite). As shown in Figure A-13, such a figure contains an inscribed circle with its center on the diagonal AC, as well as the angle bisectors of the angles in B and D, which establishes the circle.

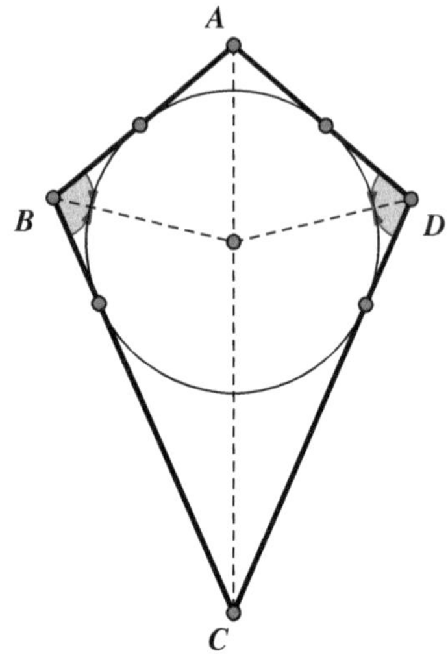

Figure A-13

We can then assume that $AB > DA$, as shown in Figure A-14. In this case, we also have $BC = AB + CD - DA > AB + CD - AB = CD$. It then follows that there exist point X on segment AB and point Y on segment BC such that $AX = DA$ and $CY = CD$. We now note that $BX = AB - AX = AB - DA = (BC + DA - CD) - DA = BC - CD = BC - CY = BY$, and it follows that the three triangles AXD, BYX, and CDY are all isosceles, with $AX = AD$, $BY = BX$, and $CD = CY$, respectively. This means that the angle bisectors of $\angle A$, $\angle B$, and $\angle C$ of the quadrilateral $ABCD$ are also the bisectors of the segments DX, XY, and YD, respectively, as these are the bases of the respective isosceles triangles. These three lines intersect in a common point, the circumcenter of triangle DXY, and we see that the angle bisectors of $\angle A$, $\angle B$, and $\angle C$ of the quadrilateral $ABCD$ intersect in a common point O. This common point, therefore, has the same distance to each of the sides DA, AB, BC, and CD of the quadrilateral, and there must exist a circle with center O that is tangent to all four sides of the quadrilateral, which is what we set out to prove.

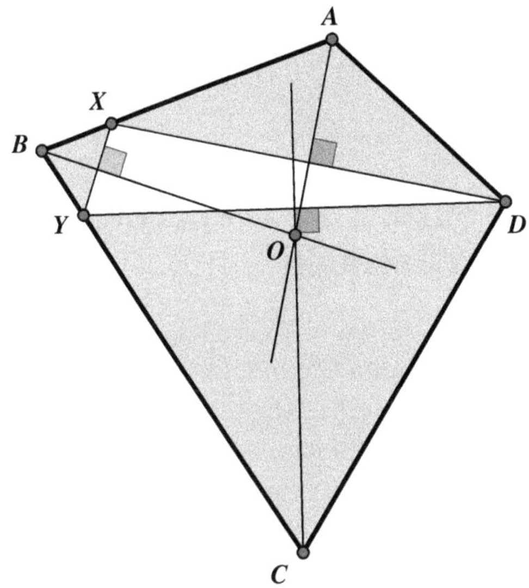

Figure A-14

Explaining Deceptive Geometry

The procedure used to solve this problem is not taught in school and requires some logical thinking. Notice that each of the two right triangles in Figure A-15 can be partitioned into congruent right triangles. On the left, the square is comprised of two of the four right triangles. Therefore, the area of the square is $\frac{2}{4}=\frac{1}{2}$ of the area of the

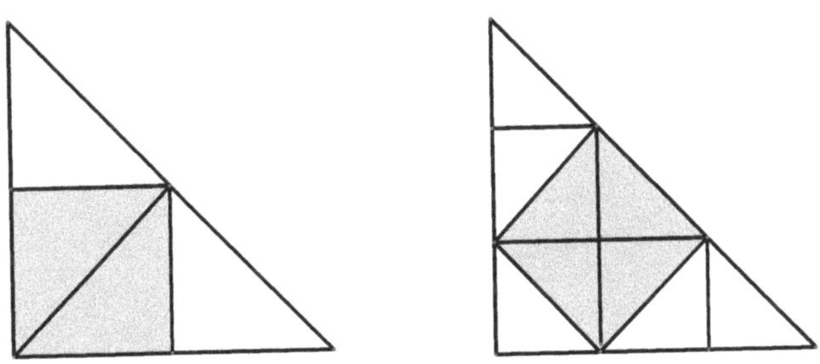

Figure A-15

right triangle. On the right side, the right triangle is partitioned into nine congruent right triangles, where the square is comprised of four of the nine right triangles. Therefore, the area of the square is $\frac{4}{9}$ of the area of the large right triangle, which is less than $\frac{1}{2}$. Thus, the larger area square is the one on the left, which is half the area of the right triangle.

Justifying the Surprising Properties of the Reuleaux Triangle

As shown in Figure A-16, we draw a tangent to the arc of the circle with center A at point P, and the distance from point P to the circle's center A is equal to the radius of the circle. This radius is equal to the lengths $b = AB = CA$ of two sides of equilateral triangle ABC, as these are also radii of the circle. The perpendicular to AP at point A is such that $AP = b$. This is also true for the tangent to the arc of the circle with center C at point Q, as all sides of the equilateral triangle have the length b. Since an analogous argument holds for tangents of the third arc, we see that pairs of parallel tangent lines of the figure always have a distance equal to the length b of the sides of triangle ABC.

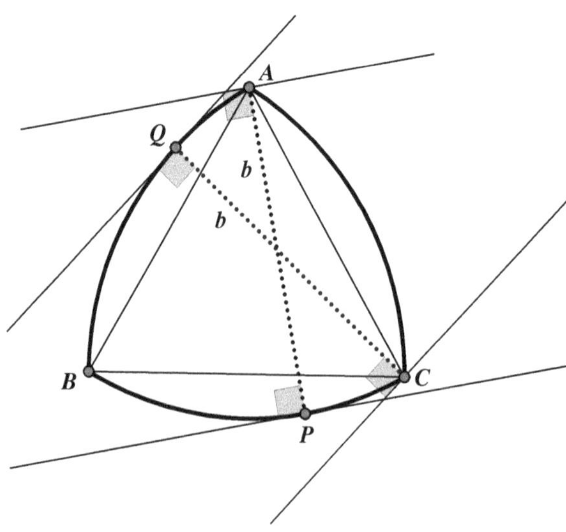

Figure A-16

A Justification for Semicircles Creating the Arbelos

In Figure A-17, points A and B are the endpoints of the large arc, and point C is the common interior endpoint of the two smaller arcs. Point O is the midpoint of AB, point P is the midpoint of AC, and point Q is the midpoint of BC. The radius of the semicircle with center P is r_1 and the radius of the semicircle with center Q is r_2. Thus, we have

$$AB = AC + CB = 2 \cdot r_1 + 2 \cdot r_2 = 2 \cdot (r_1 + r_2),$$

and the radius of circle O is $r_1 + r_2$. We know that the arc length of a semicircle is equal to π times the circle's radius. Therefore, the length of arc AC is equal to πr_1, the length of arc BC is equal to πr_2, and the length of arc AB is equal to

$$\pi (r_1 + r_2).$$

Since

$$\pi \cdot (r_1 + r_2) = \pi \cdot r_1 + \pi \cdot r_2,$$

we obtain $\widehat{AB} = \widehat{AC} + \widehat{BC}$.

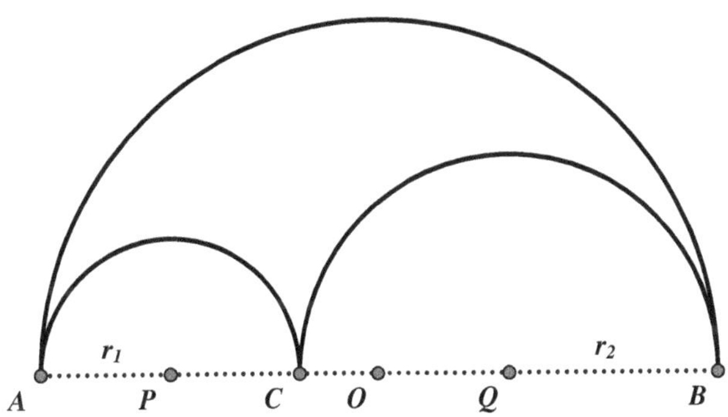

Figure A-17

A Justification for an Unexpected Arbelos Bisection

In Figure A-18, point S is the intersection of the tangents FG and HC to the two smaller semicircles. Since tangents to a circle from any external point are always of equal length, we have $SC = SF$ for circle P and $SC = SG$ for circle Q, so that $SC = SF = SG$. Next, we consider triangle AHB. Since point H lies on the circle with the diameter AC, it follows that ΔAHB is a right triangle with the altitude HC, which gives us $HC^2 = AC \cdot BC = 2r_1 \cdot 2r_2 = 4r_1r_2$. Now consider quadrilateral $FPJG$ with point J on radius QG in such a way so that $PJ \perp PF$. Since the tangent FG is perpendicular to the radii PF and GJ, it follows that $FPJG$ is a rectangle; therefore, $FG = PJ$. We also have $JQ = r_2 - r_1$ and $PQ = r_2 + r_1$. If we apply the Pythagorean theorem to the right triangle PQJ, we get

$$PJ^2 = (r_2 + r_1)^2 - (r_2 - r_1)^2 = 4r_1r_2,$$

and this gives us $FG^2 = 4r_1r_2$. Thus, $HC = FG$, and together with $SC = SF = SG$, we have $SC = SH$.

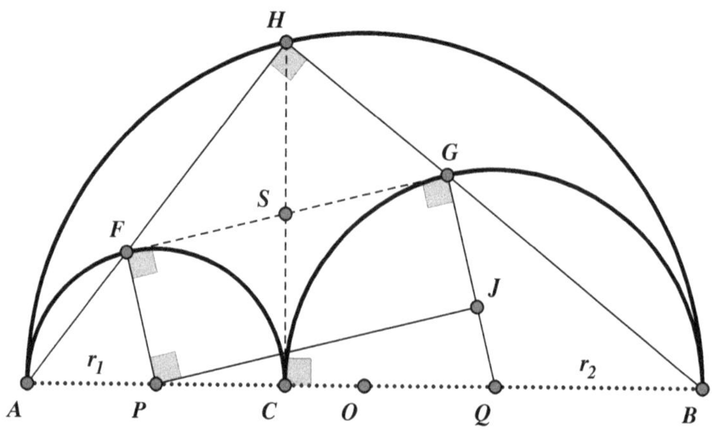

Figure A-18

A Justification for More on the Common Arbelos Bisection Point

We showed that $SC = SF = SG = SH$, so that point S is the center of a circle containing the four points F, C, G, and H, as shown in Figure A-19. The area of the arbelos can be found by taking the area of the largest semicircle and subtracting the areas of the two smaller semicircles. Since we have $AB = 2r_1 + 2r_2$, the radius of the large semicircle is equal to $r_1 + r_2$, and the area of arbelos is then equal to

$$\frac{\pi(r_1+r_2)^2}{2} - \left(\frac{\pi r_1^2}{2} + \frac{\pi r_2^2}{2}\right) = \frac{\pi}{2}\cdot\left((r_1+r_2)^2 - r_1^2 - r_2^2\right) = \frac{\pi}{2}\cdot 2r_1r_2 = \pi\cdot r_1r_2.$$

As we saw in the previous proof, the diameter FG of the circle with center S is equal to $2\sqrt{r_1r_2}$, and its area is therefore equal to πr_1r_2, which is the same as the area of the arbelos.

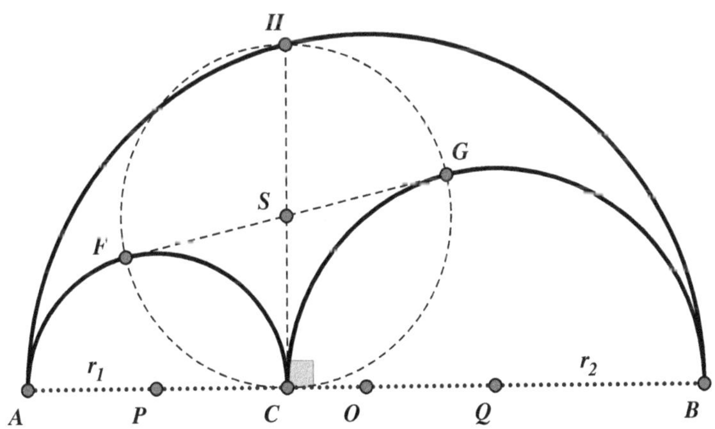

Figure A-19

A Justification for the Curious Relationship Between the Orthocenter and Circumcenter

In Figure A-20, *AD*, *BE*, and *CF* are the altitudes of triangle *ABC*, with point *D* on *BC*, point *E* on *CA*, and point *F* on *AB*. Point *P* is diametrically opposite to *B* on the circumcircle of ΔABC. Since *BP* is a diameter of circle *O*, we have $PC \perp BC$ and $PA \perp AB$. Since *AD* and *CF* are altitudes in triangle *ABC*, we also have $AH \perp BC$ and $CH \perp AB$, which gives us $AH \parallel PC$ and $CH \parallel PA$. We see that *AHCP* is a parallelogram, and thus, $AH = PC$. The perpendicular from point *O* to *BC* intersects *BC* at its midpoint *M*. In triangle *PBC*, line *OM* thus joins the midpoints of sides *PB* and *BC* and is therefore one-half the length of *PC*, so that $OM = \frac{1}{2}AH$.

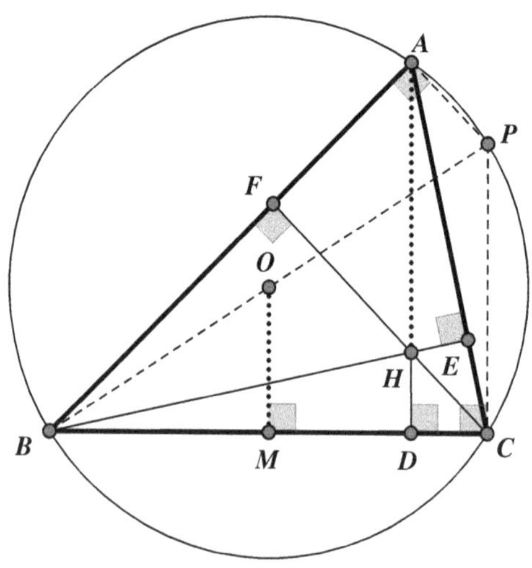

Figure A-20

The justification for the concurrency of the orthocenter and the circumcenter of triangle *ABC* is seen in Figure A-21. We begin by noting that *COQ* is a diameter in the circumcircle of triangle *ABC*.

Our initial goal here is to prove that quadrilateral *HAQB* is a parallelogram. Since *CQ* is a diameter of circle *O*, we have $\angle BQA = 180° - \angle ACB$. Also, since *BH* is an altitude in triangle *ABC* and the angles subtended on the arc of the circle are equal, we have

$$\begin{aligned}\angle HBQ &= \angle HBA + \angle ABQ = (90° - \angle BAC) + \angle ACQ \\ &= (90° - \angle BAC) + (90° - \angle CQA) = (90° - \angle BAC) + (90° - \angle CBA) \\ &= 180° - \angle BAC - \angle CBA = \angle ACB.\end{aligned}$$

We can determine that $\angle QAH = \angle ACB$ in an analogous way. With $\angle HBQ = \angle QAH = \angle ACB$, and since the opposite angles of the cyclic quadrilateral *AQBC* are supplementary, we have $\angle BQA = 180° - \angle ACB$. The angle *HBQ* is therefore supplementary to angle *BQA*, and thus, $AQ \parallel HB$ so that quadrilateral *HAQB* is a parallelogram. This means that its diagonals *AB* and *HQ* bisect each other in the midpoint *N* of *AB*. Point *Q* is, therefore, also the point *P* in which *HN* intersects circle *O*, which is what we set out to prove.

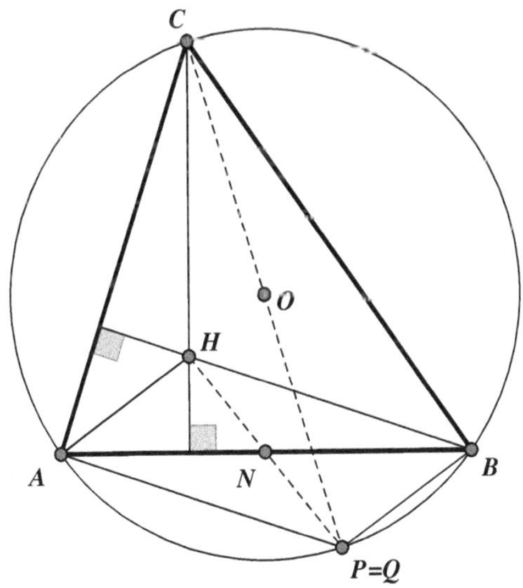

Figure A-21

Various Justifications of Morley's Theorem

Proof 1

Let D be the intersection point of the two angle trisectors of B and C adjacent to side BC. Let the other two trisectors intersect at point P, as shown in Figure A-22. Then point D is clearly the incenter of ΔPBC and PD is the bisector of $\angle BPC$. From point D two lines are constructed to intersect CP and BP at points E and F, respectively, so that $\angle PDE = \angle PFD = 30°$, which is shown in Figure A-22. We then have $\Delta PDE \cong \Delta PDF$ so that $DE = DF$. Because $\angle EDF = 60°$, we can conclude that ΔDEF is equilateral and $PD \perp EF$, as shown in Figure A-23.

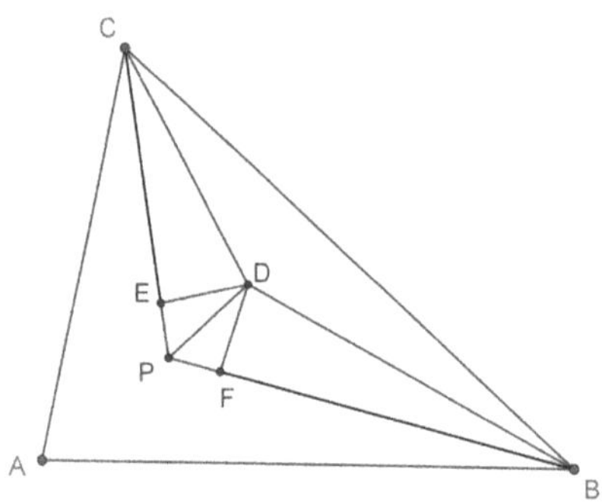

Figure A-22

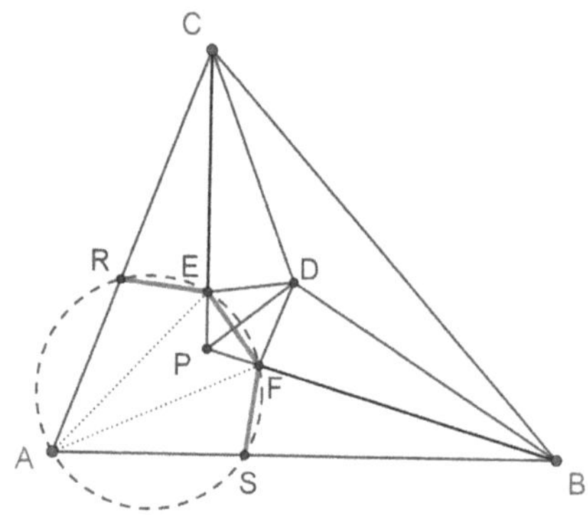

Figure A-23

It remains to be shown that AE and AF are the trisectors of $\angle BAC$. Since CP and BP are bisectors of angle ACD and angle ABD, respectively, when we reflect D in CP and in BP, we get point R on side AC and point S on side AB, respectively. The quadrilateral $REFS$ (which can be seen in Figure A-23, or for more clarity see Figure A-24) consists of three equal line segments, RE, EF, and FS, which are each equal to the side length of the equilateral ΔDEF, shown in Figure A-23, creating equal angles at E and F, which are $\angle PFS = \angle PFD = \angle PED = \angle PER$ and $\angle PFE = \angle PEF$. By addition, this results in $\angle EFS = \angle REF$. In addition to establishing that these two angles are equal, we can also calculate their measure as follows:

$$\angle BPC = \underbrace{180^\circ}_{\angle A+\angle B+\angle C} - \frac{2}{3}\angle B - \frac{2}{3}\angle C = \underbrace{\frac{1}{3}(\angle A + \angle B + \angle C)}_{60^\circ} + \frac{2}{3}\angle A = 60^\circ + \frac{2}{3}\angle A,$$

so that $\angle PFE = 90^\circ - \frac{1}{2}\angle BPC = 60^\circ - \frac{1}{3}\angle A$. Then $\angle PFS = \angle PFD = 60^\circ + \angle PFE = 120^\circ - \frac{1}{3}\angle A$, and $\angle EFS = \angle PFE + \angle PFS = 180^\circ - \frac{2}{3}\angle A$. We would get the same result for $\angle REF$, but it is not necessary to do the calculation again, because we know from the above that $\angle EFS = \angle REF$.

Hence, *REFS* is an isosceles trapezoid, as shown in Figure A-23, and thus is *cyclic*. Moreover, *REFS* has the supplementary angles $\angle REF = 180^\circ - \frac{2}{3}\angle A = \angle EFS$, and $\angle ERS = \frac{2}{3}\angle A = \angle FSA$. In trapezoid *REFS*, the diagonals are angle bisectors of $\angle R$ and $\angle S$, since $\angle RFE = \angle FRE$ resulting from isosceles ΔRFE, and $\angle RFE = \angle FRS$ resulting from $RS \parallel EF$. Hence, we have

$$\angle RFS = \angle EFS - \underbrace{\angle RFE}_{\frac{1}{2}\angle ERS} = 180^\circ - \frac{2}{3}\angle A - \frac{1}{2}\cdot\frac{2}{3}\angle A = 180^\circ - \angle A.$$

This establishes that $\angle RFS$ is supplementary to $\angle A$, and it therefore follows that point *A* lies on the circumcircle of *REFS*. Since the chords $RE = EF = SF$ are equal, they determine equal inscribed angles $\angle RAE = \angle EAF = \angle FAS$, so that we have *AE* and *AF* as angle trisectors of $\angle BAC$. Thus, the trisectors of the three angles of triangle *ABC* form equilateral triangle *DEF*.

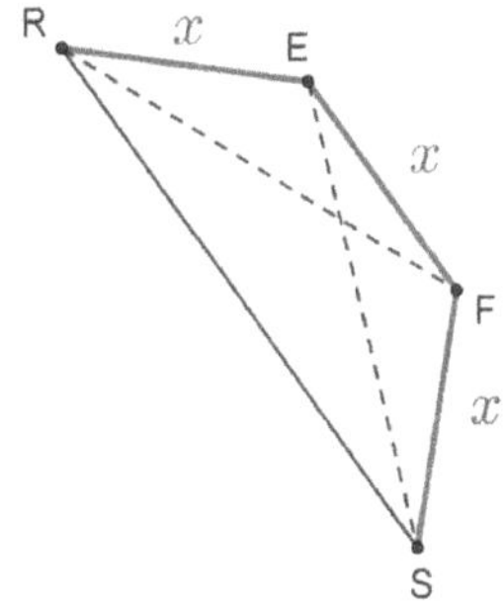

Figure A-24

Proof 2

We begin with a randomly drawn triangle *ABC*, shown in Figure A-25, and then define $\frac{1}{3}\angle A=\alpha$, $\frac{1}{3}\angle B=\beta$, and $\frac{1}{3}\angle C=\gamma$. We then have $\alpha+\beta+\gamma=60°$. To simplify matters, we will denote with α' the angle $\alpha+60°$ (analogously for β, γ).

We will employ a different proof strategy here. We will not begin with ΔABC and then show that ΔDEF is equilateral. Instead, we will be *working backwards*, that is, we will start with an equilateral $\Delta D'E'F'$ and construct $\Delta A'B'C'$, which will prove to be similar to ΔABC. Moreover, we will see that in $\Delta A'B'C'$ the points D', E', F' are nothing less than the intersection points of the angle trisectors of $\Delta A'B'C'$. This will allow us, finally, to conclude the similarity $\Delta D'E'F' \sim \Delta DEF$, which means that ΔDEF is equilateral, as we sought to show. This kind of working backwards here is a clever principle.

Now let us describe this path of reasoning in more detail. As illustrated in Figure A-25, we begin with the equilateral $\Delta D'E'F'$ and erect on its sides three outward triangles with the vertices A', B', C'. We will now show that $\angle F'A'B'=\alpha$ and $\angle F'B'A'=\beta$. To do this, we consider the perpendiculars $F'X$ and $F'Y$ to $B'D'$ and $A'E'$, respectively. Since $\angle A'E'F'=\angle F'D'B'=\gamma'$ and $E'F'=D'F'$, right triangles $\Delta F'D'X \cong \Delta F'E'Y$, and we then have $F'X=F'Y=d$.

First, we can calculate $\angle A'F'B'=360°-\beta'-60°-\alpha'=180°-(\alpha+\beta)$, which we denote as equation (*).

We then consider the perpendicular distance h from F' to $A'B'$, measuring this distance along line segment $F'Z$. If $h<d$, then the following holds (note that in a right triangle with a fixed hypotenuse, increasing an angle yields longer opposite legs and vice versa): $\angle B'A'F'<\alpha$ and $\angle F'B'A'<\beta$, and thus, $\angle A'F'B'=180°-\angle B'A'F'-\angle F'B'A'>180°-(\alpha+\beta)$, which contradicts equation (*).

On the other hand, if $h>d$, it follows that $\angle B'A'F'>\alpha$ and $\angle F'B'A'>\beta$, and thus, $\angle A'F'B'<180°-(\alpha+\beta)$, again a contradiction to equation (*). Hence, it follows $h=d$, $\angle F'A'B'=\alpha$, and $\angle F'B'A'=\beta$.

In an analogous way, we can show that $\angle C'B'D'=\beta$, $\angle D'C'B'=\angle A'C'E'=\gamma$, and $\angle E'A'C'=\alpha$. We see that $A'E'$ and $A'F'$ trisect $\angle B'A'C'$, $B'F'$ and $B'D'$ trisect $\angle C'B'A'$, and $C'D'$ and $C'E'$ trisect $\angle A'C'B'$. Since $\angle B'A'C'=3\alpha$, $\angle C'B'A'=3\beta$, and $\angle A'C'B'=3\gamma$, we know that $\Delta A'B'C'$ is similar to the originally given ΔABC. This similarity extends to all constructions, and the trisections of the interior angles of ΔABC therefore yield an internal ΔDEF similar to $\Delta D'E'F'$, which must also be equilateral.

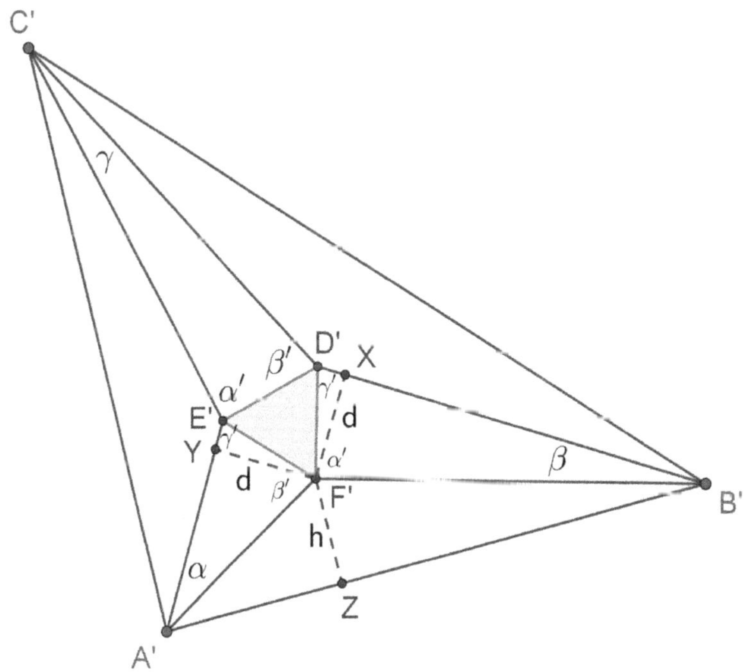

Figure A-25

Justifying Napoleon's Theorem

To prove that triangle KLM is equilateral, using Figure A-26 we will show that the sides are proportional to the equal lengths AE, BF, and CD. We first let N denote the midpoint of AC. Since MA bisects $\angle CAF$, we know that $\angle NAM = \angle CAM = 30°$. We also know that point M is the center of triangle ACF and, therefore, $MN \perp AC$, which enables us to identify triangle ANM as a 30°–60°–90° triangle. As a result, in triangle ANM we find that

$$\frac{AN}{AM} = \frac{\sqrt{3}}{2},$$

or

$$\frac{AC}{AM} = \frac{2\sqrt{3}}{2} = \sqrt{3}.$$

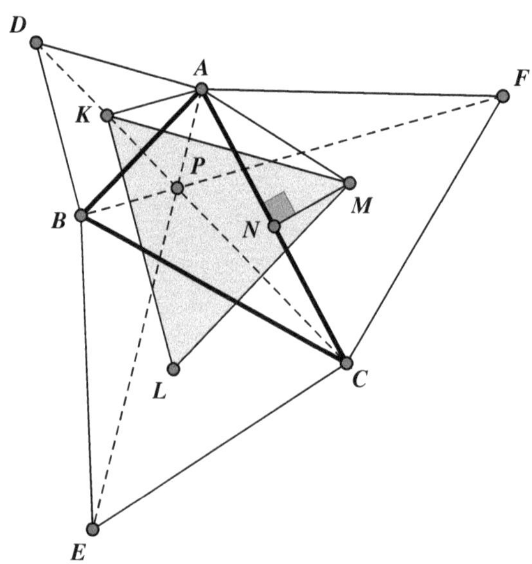

Figure A-26

Analogously, in triangle DBA, we have $\frac{AB}{AK}=\sqrt{3}$. We also know that $\angle KAM = \angle KAB + \angle BAM = 30° + \angle BAM = \angle BAM + \angle MAF = \angle BAF$, which then allows us to establish that triangles AKM and ABF are similar. It then follows that

$$\frac{BF}{KM}=\frac{AB}{AK}=\sqrt{3}.$$

Furthermore, in the same fashion we can show that

$$\frac{DC}{KL}=\sqrt{3},$$

and

$$\frac{AE}{ML}=\sqrt{3}.$$

We now have

$$\frac{BF}{KM}=\frac{DC}{LK}=\frac{AE}{ML}.$$

However, since we know that $BF = DC = AE$, we can conclude that $KM = LK = ML$. Therefore, triangle KLM is equilateral.

The equality of the lines joining the three vertices of triangle ABC with the remote vertices of the equilateral triangles can easily be justified by showing that the two triangles ABE and DBC are congruent, both of which are shown shaded in Figure A-27. The fact that the triangles ADB and BEC are equilateral gives us $BA = BD$, $BE = BC$, and $\angle CBD = \angle CBA + \angle ABD = \angle CBA + 60° = \angle CBA + \angle EBC = \angle EBA$, and triangles ABE and DBC are, therefore, congruent. From this, we immediately obtain $AE = CD$. Similarly, we can also show in an analogous way that triangles ABF and ADC are congruent, which then gives us the equality $AE = BF = CD$. This could also have been done using ΔCFB and ΔCAE.

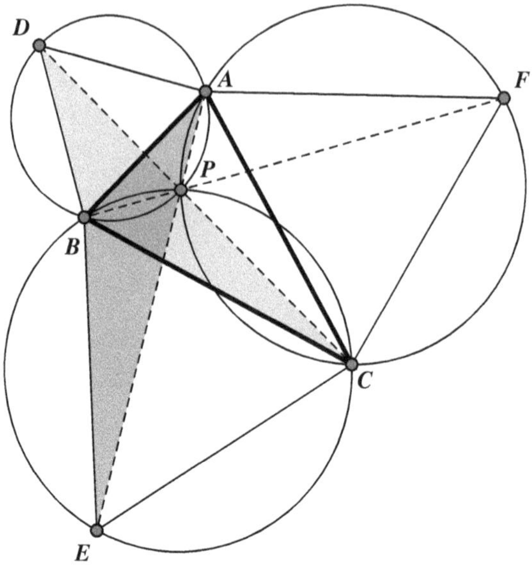

Figure A-27

In order to show that these three lines are concurrent, we will have to take a slightly closer look at the situation. Letting P denote the point in which AE and CD intersect, we note that $\angle BAP = \angle BAE = \angle BDC = \angle BDP$ because the triangles ABE and DBC are congruent, as can be seen in Figure A-27. This means that points P, A, D, and B lie on a common circle, and we therefore have $\angle APB = 180° - \angle BDA = 180° - 60° = 120°$. Similarly, we also have $\angle PCB = \angle DCB = \angle AEB = \angle PEB$, which means that points P, B, E, and C also lie on a common circle, and we then also have $\angle BPC = 180° - \angle CEB = 180° - 60° = 120°$. It thus follows that $\angle CPA = 360° - \angle APB - \angle BPC = 360° - 120° - 120° = 120°$, and since $\angle AEC = 60°$, points A, P, C, and F also lie on a common circle. This means that $\angle FPA = \angle FCA = 60°$, from which we obtain $\angle APB + \angle FPA = 120° + 60° = 180°$. Point P therefore also lies on BF, and we then have P as a common point of AE, BF, and CD.

We have already shown $\angle APB = \angle BPC = \angle CPA = 120°$. What remains to be shown is that the sum of the distances from P to the

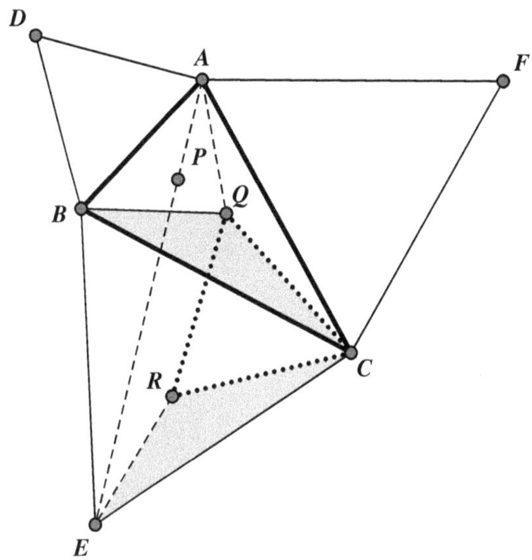

Figure A-28

three triangle vertices $PA+PB+PC$ is minimal among all points in the interior of the triangle. To show this, we choose any point Q, not necessarily in the interior of the triangle, as shown in Figure A-28.

We wish to discover a property of the sum of the distances from Q to the triangle vertices, or in other words, the expression $QA+QB+QC$. To facilitate a proof, we add an equilateral triangle CQR to Figure A-28. Since the sides of this triangle are of equal length, we have $QC = QR$. Also, we note that triangles QBC and REC are congruent, since we have $QC=RC$, $BC=EC$, and $\angle QCB = \angle QCR - \angle BCR = 60° - \angle BCR = \angle BCE - \angle BCR = \angle RCE$. This implies $QB=RE$. We see that $QA+QB+QC=AQ+QR+RE$. If Q does not lie on the line AE, the assumption that this sum is minimal is obviously not correct, since a choice of Q on AE results in a smaller value of the sum. We see that point P, for which the sum $PA+PB+PC$ is minimal, must certainly lie on AE. However, we can argue analogously for BF and CD, and we see that the common point P of AE, BF, and CD we discovered earlier has the property of being the minimal distance point, as we had set out to prove.

A Justification for Establishing a Concurrency Point of Circles

In Figure A-29, we have P as the intersection of AE and CD $\angle APB = \angle BPC = 120°$. Since $\angle BDA = \angle CEA = 60°$, this means that P is an intersection point of the circumcircle of triangles ADB and BEC. Since $\angle CPA = 360° - \angle APB - \angle BPC = 360° - 120° - 120° = 120°$ and $\angle AFC = 60° = 180° - 120° = 180° - \angle CPA$, we then see that P also lies on the circumcircle of triangle CFA. Thus, P lies on all three circumcircles, which proves the circle concurrency.

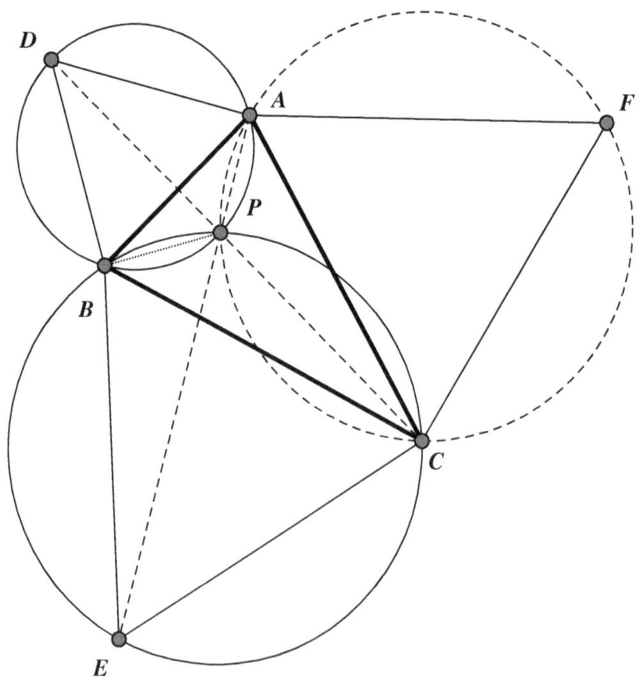

Figure A-29

Establishing the Miquel Point

Consider the situation where point *M* is inside triangle *ABC*, as shown in Figure A-30. Points *D*, *E*, and *F* are any points on the sides of *AC*, *BC*, and *AB*, respectively, of triangle *ABC*. We will consider the circles *Q* and *R*, determined by the points *F*, *B*, *E* and *D*, *C*, *E*, respectively, to meet at point *M*. We will now draw line segments *MF*, *ME*, and *MD*.

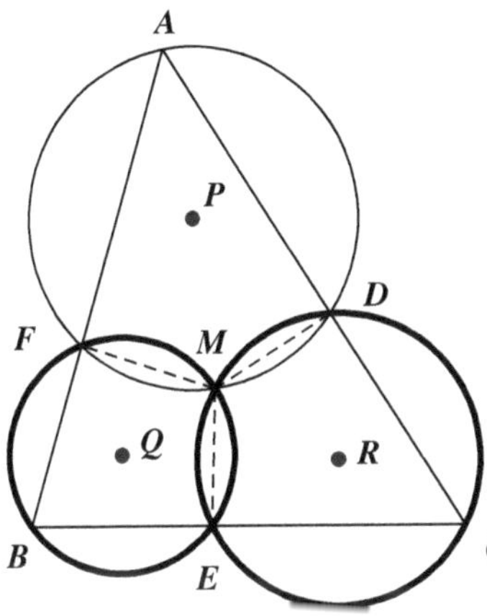

Figure A-30

Because quadrilateral *BFME* is cyclic, and the opposite angles of a cyclic quadrilateral are supplementary, $\angle FME = 180° - \angle B$. Similarly, quadrilateral *CDME* is also cyclic; therefore, $\angle DME = 180° - \angle C$. When we add these two angles, we get $\angle FME + \angle DME = 360° - (\angle B + \angle C)$, or written another way, $\angle B + \angle C = 360° - \angle FME + \angle DME = \angle DMF$. However, in triangle *ABC* we notice that $\angle B + \angle C = 180° - \angle A$. It then follows that $\angle DMF = 180° - \angle A$, which allows us to conclude that quadrilateral *AFMD* is cyclic. Therefore, the circle containing points *D*, *A*, and *F* contains a point common to the other two circles, namely, point *M*.

Justifying Menelaus' Theorem

In Figure A-31, we have triangle *ABC* with three points, *D*, *E*, and *F*, on sides *BC* (extended), *CA*, and *AB*, respectively. Menelaus' theorem states that the three points *D*, *E*, and *F* are collinear if, and only if, the alternate segments they determine on the sides have equal products. Expressed as an equation, this can be written as $AF \cdot BD \cdot CE = AE \cdot BF \cdot CD$.

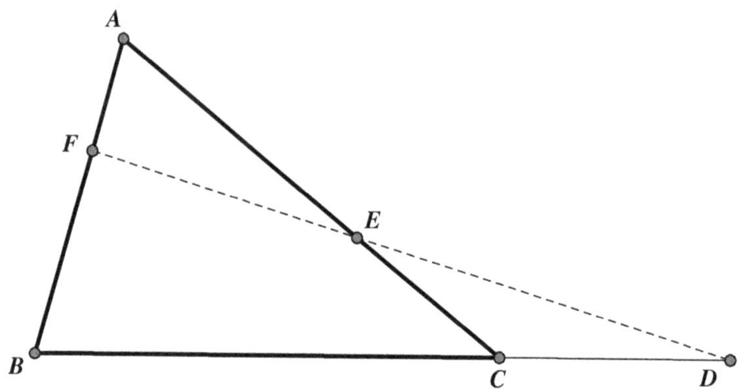

Figure A-31

We will divide the proof of this into several steps. First of all, we assume that *D*, *E*, and *F* are collinear and show that $AF \cdot BD \cdot CE = AE \cdot BF \cdot CD$ must then follow. In Figure A-32, we draw a line containing point *C*, parallel to *AB*, and intersecting *DEF* at point *P*.

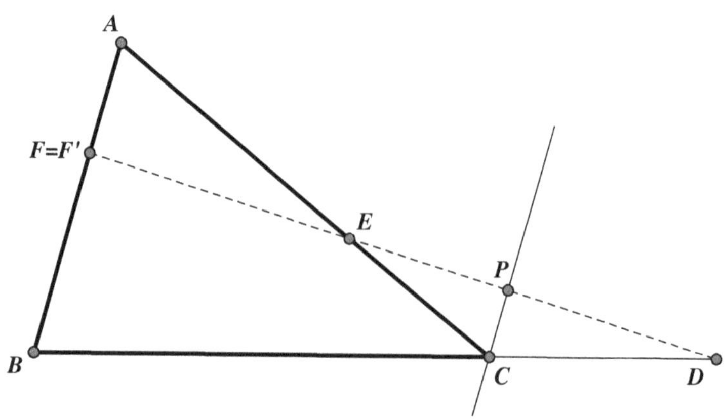

Figure A-32

Since triangles PCD and FBD are similar, we have

$$\frac{CP}{BF} = \frac{CD}{BD},$$

or

$$CP = \frac{BF \cdot CD}{BD}.$$

Furthermore, triangles PEC and FEA are also similar, and we therefore have

$$\frac{CP}{AF} = \frac{CE}{AE},$$

or

$$CP = \frac{AF \cdot CE}{AE}.$$

This gives us

$$\frac{BF \cdot CD}{BD} = \frac{AF \cdot CE}{AE},$$

which is equivalent to $AF \cdot BD \cdot CE = AE \cdot BF \cdot CD$. This proves the first part of Menelaus' theorem.

Next, we assume that $AF \cdot BD \cdot CE = AE \cdot BF \cdot CD$, and show that points D, E, and F must be collinear. We let the intersection point of AB and DE be the point F'. Then we have to prove $F' = F$. Since points D, E, and F' are collinear, we have just shown that $AF' \cdot BD \cdot CE = AE \cdot BF' \cdot CD$ holds, which we can also write as

$$\frac{AE}{CE} \cdot \frac{BF'}{AF'} \cdot \frac{CD}{BD} = 1.$$

Since we are assuming that $AF \cdot BD \cdot CE = AE \cdot BF \cdot CD$, which we can also write as

$$\frac{AE}{CE} \cdot \frac{BF}{AF} \cdot \frac{CD}{BD} = 1,$$

we therefore have

$$\frac{BF'}{AF'} = \frac{BF}{AF},$$

and so we have shown that $F = F'$.

As a last step, we now consider the configuration illustrated in Figure A-33.

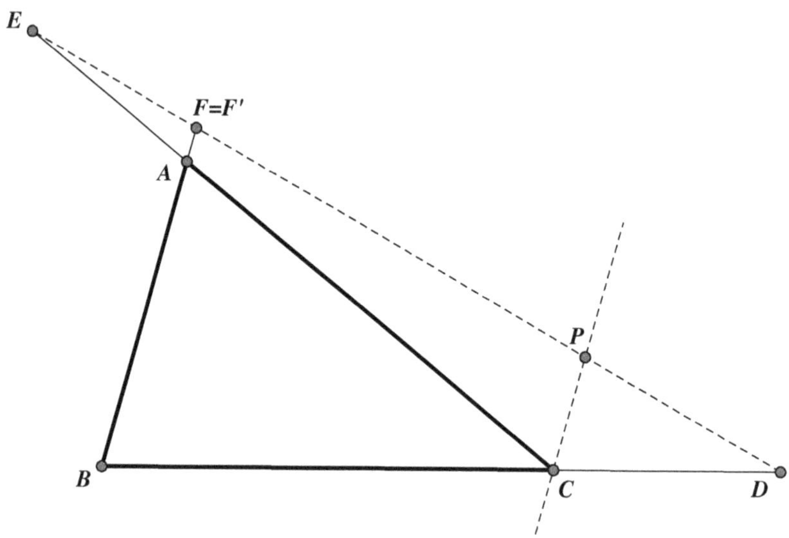

Figure A-33

Here, we have triangle ABC with points D, E, and F on the extended sides BC, CA, and AB, respectively. In this situation, we can also show that the three points D, E, and F are collinear if, and only if, $AF \cdot BD \cdot CE = AE \cdot BF \cdot CD$. The proof in this case is identical to that previously shown.

Menelaus' theorem can be stated in a more general fashion. We are given a triangle ABC. Points D, E, and F lie on the lines BC, CA and AB, respectively. These points are collinear if, and only if, $AF \cdot BD \cdot CE = AE \cdot BF \cdot CD$, or

$$\frac{AE}{CE} \cdot \frac{BF}{AF} \cdot \frac{CD}{BD} = 1.$$

Justifying Simson's Theorem

We begin by drawing *PA*, *PB*, and *PC* in Figure A-34. Since *A*, *B*, *C*, and *P* lie on a common circle, we have $\angle ABP = \angle ACP$. Right triangles *ZPB* and *YPC* are therefore similar, and we have

$$\frac{BZ}{CY} = \frac{PZ}{PY}.$$

Similarly, $\angle PAB = \angle PCB$ gives us similar right triangles *XCP* and *ZAP*, so that

$$\frac{CX}{AZ} = \frac{PX}{PZ}.$$

Since $\angle CBP$ and $\angle PAC$ are opposite angles of an inscribed (cyclic) quadrilateral, they are supplementary. However, $\angle YAP$ is also supplementary to $\angle PAC$. This means that right triangles *XPB* and *YPA* are also similar, giving us

$$\frac{AY}{BX} = \frac{PY}{PX}.$$

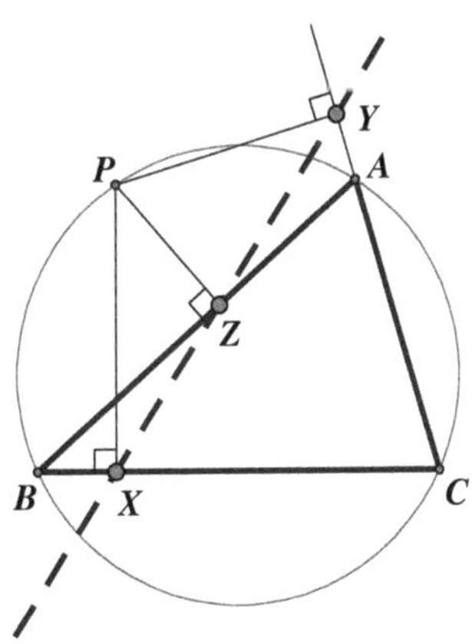

Figure A-34

Multiplying these identities, we obtain

$$\frac{BZ}{CY}\cdot\frac{CX}{AZ}\cdot\frac{AY}{BX}=\frac{PZ}{PY}\cdot\frac{PX}{PZ}\cdot\frac{PY}{PX}=1.$$

Thus, by Menelaus' theorem (which we proved above), the points X, Y, and Z are collinear. These three points determine the *Simson line* of ΔABC with respect to point P.

The Strange Characteristic of a Hexagon

Now we can apply Menelaus' theorem to prove Pascal's theorem. In Figure A-35, A, B, C, D, E, and F are points on the circle c, and $P = BC \cap DE$, $Q = BC \cap EF$ and $R = CD \cap FA$ are defined as required.

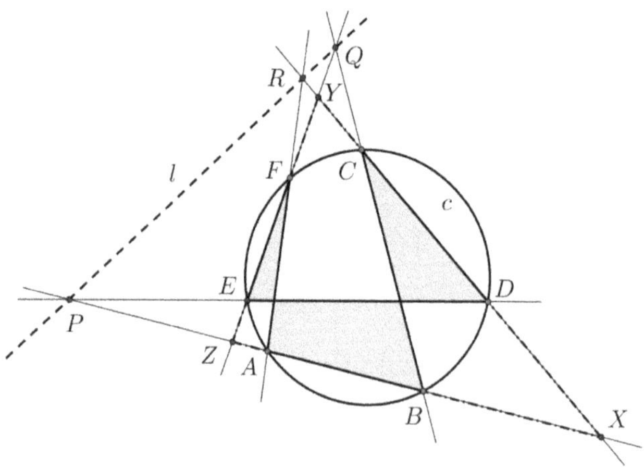

Figure A-35

We now also define three new points, namely, $X = AB \cap CD$, $Y = CD \cap EF$, and $Z = EF \cap AB$.

Applying Menelaus' theorem to line ED with respect to triangle XYZ gives us

$$\frac{XD}{DY}\cdot\frac{YR}{EZ}\cdot\frac{ZP}{PX}=-1.$$

Applying Menelaus' theorem to line BC with respect to triangle XYZ gives us

$$\frac{XC}{CY}\cdot\frac{YQ}{QZ}\cdot\frac{ZB}{BX}=-1.$$

And then applying Menelaus' theorem to line FA with respect to triangle XYZ gives us

$$\frac{XR}{RY}\cdot\frac{YF}{FZ}\cdot\frac{ZA}{AX}=-1.$$

Multiplying these three equations gives us

$$\left(\frac{XD}{DY}\cdot\frac{YE}{EZ}\cdot\frac{ZP}{PX}\right)\left(\frac{XC}{CY}\cdot\frac{YQ}{QZ}\cdot\frac{ZB}{BX}\right)\left(\frac{XR}{RY}\cdot\frac{YF}{FZ}\cdot\frac{ZA}{AX}\right)=-1,$$

which can also be written as

$$\frac{XR}{RY}\cdot\frac{YQ}{QZ}\cdot\frac{ZP}{PX}\cdot\frac{XD\cdot XC}{AX\cdot BX}\cdot\frac{YF\cdot YE}{CY\cdot DY}\cdot\frac{ZA\cdot ZB}{EZ\cdot FZ}=-1.$$

Due to the power of a point theorem,[1] we have $XD \cdot XC = AX \cdot BX$, $YF \cdot YE = CY \cdot DY$, and $ZA \cdot ZB = EZ \cdot FZ$, which means that the above equation reduces to

$$\frac{XR}{RY}\cdot\frac{YQ}{QZ}\cdot\frac{ZP}{PX}\cdot 1\cdot 1\cdot 1=-1 \Leftrightarrow \frac{XR}{RY}\cdot\frac{YQ}{QZ}\cdot\frac{ZP}{PX}=-1.$$

This is exactly the property required for Menelaus' theorem. Therefore, we see that P, Q, and R lie on a common line l, as claimed.

---

[1] A circle partitions two secants drawn from the same external point such that the product of one secant and its external segment is equal to the product of the other secant and its external segment.

Index

www.ingramcontent.com/pod-product-compliance
Lightning Source LLC
LaVergne TN
LVHW070827220926
848486LV00028B/83
9789819830411